METHODS OF NEUROCHEMISTRY
Volume 5

Edited by RAINER FRIED

Department of Biochemistry
Creighton University
Medical School
Omaha, Nebraska

Springer Science+Business Media, LLC 1973

ISBN 978-1-4899-6186-0   ISBN 978-1-4899-6375-8 (eBook)
DOI 10.1007/978-1-4899-6375-8

LIBRARY OF CONGRESS CATALOG CARD NUMBER 75-134782

CONTRIBUTORS TO THIS VOLUME

Ata A. Abdel-Latif, Department of Cell and Molecular Biology, Medical College of Georgia, Augusta, Georgia

E. Gutmann, Institute of Physiology, Czechoslovak Academy of Sciences, Prague, Czechoslovakia

Richard H. Himes, Department of Physiology and Cell Biology, and Department of Biochemistry, University of Kansas, Lawrence, Kansas

Serge B. Melancon, Department of Pediatrics, Northwestern University School of Medicine, and Division of Genetics, The Children's Memorial Hospital, Chicago, Illinois

Henry L. Nadler, Department of Pediatrics, Northwestern University School of Medicine, and Division of Genetics, The Children's Memorial Hospital, Chicago, Illinois

Dianna Ammons Redburn, Department of Physiology and Cell Biology, and Department of Biochemistry, University of Kansas, Lawrence, Kansas

Frederick E. Samson, Jr., Department of Physiology and Cell Biology, and Department of Biochemistry, University of Kansas, Lawrence, Kansas

Lewis S. Seiden, University of Chicago, Departments of Pharmacology and Psychiatry, Chicago, Illinois

CONTENTS

CONTENTS OF VOLUME 4

*Present address: Fresno Community Hospital, Fresno, California

Chapter 1

PRENATAL DIAGNOSIS OF GENETIC DISORDERS
LEADING TO MENTAL RETARDATION

Serge B. Melancon and Henry L. Nadler
Department of Pediatrics
Northwestern University School of Medicine

and

Division of Genetics
The Children's Memorial Hospital
Chicago, Illinois

1

I. INTRODUCTION

The prenatal detection of genetic disorders has become an area of increasing importance during the past few years. The development of reliable screening procedures for the early detection of familial disorders leading to mental retardation has stimulated the demand for an early assessment of fetal integrity in "high risk" pregnancies. The acceptance of transabdominal amniocentesis as a safe procedure and advances in tissue culture techniques as well as micromethods for biochemical assays and chromosomal analysis have all contributed to the rapid advances in this field.

The indications, accuracy, and reliability of the procedures used in prenatal detection of chromosomal aberrations are clearly established. In contrast, the experience in management of patients at "high risk" for a familial biochemical disorder is limited and therefore the procedures should be restricted to a research center in which expertise in that specific disorder has been gained.

A number of excellent reviews have been published recently which provide guidelines as to the indications, risks, and complications of amniocentesis and related techniques used in prenatal genetic diagnosis (1-4). This chapter is designed to provide the reader with detailed methodology, as currently used in our laboratory, to perform prenatal genetic diagnosis.

II. AMNIOTIC FLUID

A. Composition

The volume of amniotic fluid at 10 weeks of gestation, averages 30 ml and gradually increases to 350 ml at 20 weeks (5-7). Amniotic fluid is composed of 98 to 99% water with 1 to 2% solids. The inorganic constituents of amniotic fluid are quite similar to extracellular fluid, and the solids are equally divided between organic solids and proteins. Excellent reviews of the composition of amniotic fluid have recently been published (8, 9).

The concentration of electrolytes in amniotic fluid prior to 20 weeks of pregnancy is essentially that found in maternal serum (10, 11). During the latter half of pregnancy, the amniotic fluid becomes increasingly hypotonic. The proteins of amniotic fluid have been investigated and although considerable variations occur, useful ranges have been established (Table 1). Conflicting reports have suggested that the proteins are of maternal origin (16, 17) while others present evidence of their fetal origin (15, 18).

TABLE 1

Protein Components of Early Amniotic Fluid (12-15)

Month of gestation	Total protein, g/liter		Albumin	Globulins			
	Mean	Range		$\alpha 1$	$\alpha 2$	β	γ
2 to 4	3.9	1.2-6.2	71.3%	10.9%	6.1%	7.4%	4.1%
5 to 6	8.1	5.6-14.3	68.0%	13.3%	7.6%	6.7%	4.4%

Other nitrogenous constituents such as urea, creatinine, and uric acid increase late in pregnancy to reach a term value approximately twice that found in maternal serum (12, 14, 19). The amino acid composition of amniotic fluid has been studied and its relevance to prenatal diagnosis of the genetic aminoacidopathies is discussed below. Many hormones have been detected in amniotic fluid (20-33). The antenatal diagnosis of the adrenogenital syndrome has been established by measuring the amniotic fluid levels of 17-ketosteroids and pregnanetriol late in pregnancy (21, 34, 35).

A number of enzymes have been detected in amniotic fluid (3). In most instances it is impossible to determine the actual source of the enzyme. Potential sources of error include maternal blood contamination and the presence of isozymes and fetal enzymes.

Amniotic fluid levels of sugars, lipids, bilirubin, protein-bound iodine, gases, and other components have also been reviewed (8, 9) and have not as yet proved useful for the in utero detection of genetic disorders. They are, however, extremely valuable for the estimation of fetal age and maturity.

B. Amniocentesis

Transabdominal amniocentesis performed between 13 and 18 weeks of pregnancy has gained widespread acceptance as being the procedure of choice to obtain amniotic fluid specimens (1-4). Transvaginal amniocentesis carries a much higher risk of complication and is not advised early in pregnancy (36). The reliability of transabdominal amniocentesis has been well established in the management of Rh disease (39-44) and lately in obtaining material for prenatal genetic diagnosis early in pregnancy (2, 45, 46).

Transabdominal amniocentesis is performed by an experienced obstetrician as an out-patient technique after thorough explanation of the risks and with signed permission of the pregnant woman and her husband. After the patient voids, a local anesthetic (1% lidocaine) is injected into the proposed puncture site. A 22-gauge, 5-in. disposable spinal needle with stylet is inserted through the abdominal wall in the midline, directed at a right angle toward the middle of the uterine cavity. After puncture, the stylet is removed and a sterile plastic syringe is used to withdraw 10 ml of amniotic fluid, after which the needle is swiftly withdrawn. Strict aseptic conditions are observed throughout the procedure (3). Other potential approaches to obtaining fetal materials is through direct biopsy of the placenta (37) and the fetus (38). These techniques have recently been performed previous to therapeutic abortion and represent highly hazardous procedures which should be done exclusively after accurate localization of the fetus and the placenta.

C. Handling of Amniotic Fluid

The amniotic fluid is placed in either a sterile siliconized glass or polystyrene tube (17 x 100 mm, 16 ml capacity) and transported at ambient temperature to the laboratory. Shipment to a distant destination is best achieved under the same conditions. The laboratory to which the fluid is to be sent should be notified before the procedure as to date, indication, test to be performed, and method of shipment. Cell viability is preserved for as long as one week at ambient temperature under sterile conditions. It is not advised to freeze the sample if cultivation of the amniotic fluid cells is planned. Shipment in culture tubes or flasks is not recommended since part or all of the cells may attach at random to the walls and thus become difficult to recover. Proper labeling of the samples is mandatory. In the laboratory, the fluid is centrifuged at 100 x g for 10 min at room temperature in a clinical centrifuge. The supernatant is removed and either processed immediately or frozen away for later reference. The cell pellet is now available for direct analysis or cultivation.

III. AMNIOTIC FLUID CELLS

A. Origin and Utilization

Amniotic fluid cells have been shown to be derived from the amnion and fetus (47-51). The origin of the fetal cells is presumably buccal mucosa, vaginal epithelium, skin, umbilical cord, and fetal urine. The number of amniotic fluid cells have been shown to increase with gestation while the percentage of viable cells decreases.

The utilization of the cellular material found in amniotic fluid initially focused upon the technique of sex chromatin analysis for the antenatal determination of sex (52-61). The presence of sex chromatin in amniotic fluid cells has been useful for the management of pregnancies in women heterozygous for X-linked recessive disorders such as hemophilia, nephrogenic diabetes insipidous, and muscular dystrophy (2, 62, 63).

The biochemical properties of uncultured amniotic fluid cells are still being investigated. Many problems such as maternal blood cell contamination (64, 65), enzyme instability (66), and insufficient number of metabolically active amniotic fluid cells have limited the usefulness of this approach to prenatal genetic diagnosis. At the present time it is mandatory to confirm the results with enzyme assay on cultivated amniotic fluid cells from the same original sample. In addition, if the uncultured amniotic fluid cells are to be used, a control enzyme must be assayed.

An increasing number of investigators demonstrate the ability to culture amniotic fluid cells. The rate of successful cultivation of amniotic fluid cells varies greatly from one laboratory to another. A number of factors probably account for the variation of these rates. In several early studies, the amniotic fluid was obtained during the third trimester of pregnancy from women with Rh isoimmunization. Cells obtained during this period are considerably more difficult to cultivate. If only those cases in which the amniotic fluid is obtained prior to twenty weeks of pregnancy are considered, the ability to cultivate these cells is over 90%.

The methods used to cultivate amniotic fluid cells differ significantly from one another; however, success rates do not appear to be dependent upon any specific factor with the possible exception of the investigator's experience and care in handling cells.

B. Cultivation of Amniotic Fluid Cells

1. Material

   a. Growth medium: to 500 ml of Ham's F-10 medium is added 75 ml of fetal calf serum and 5 ml of an antibiotic-antimycotic mixture containing 10,000 units of penicillin, 10,000 μg streptomycin, and 250 μg fungizone per ml.
   b. Puck's saline A: to 500 ml of sterile deionized water is added 55 ml of Puck's saline A (10 x) and 1 ml of 7.5% sodium bicarbonate.
   c. Trypsin solution: a 0.25% trypsin solution is freshly prepared by diluting a 2.5% stock solution with Puck's saline A.

2. Procedure

   Ten to 15 ml of amniotic fluid is gently centrifuged at 100 x g for 10 min and the cell pellet suspended in 0.1 ml to 0.2 ml of amniotic fluid, or 100% fetal calf serum. A drop of this suspension is placed in a 35 x 10 mm (8 cm^2) Petri dish (Falcon plastic). Usually three or more Petri dishes are used depending on the size of the cell pellet. A concentrated cell suspension has a better chance of attachment and therefore cultivation. A sterile glass coverslip is placed on each drop to immobilize the amniotic fluid cells, and the dishes are placed in a CO_2 incubator at 37°C for 20 min. Two to 3 ml of growth mixture are added and the dishes incubated in 5% CO_2 at 37°C. The medium is changed every other day until a number of colonies of cells are seen under the coverslips, usually 7 to 18 days. The medium is removed from the dish, and the coverslip turned cell surface up and placed in another Falcon dish. Medium is added to both the original and the new dish. Staining procedures and chromosome analysis can be performed directly on the coverslip while the cells in the original dish are usually maintained in long term culture. When confluency is obtained in the original dish, the medium is removed, and the cells washed twice in Puck's saline A and trypsinized (0.1-0.2 ml of 0.25% trypsin until the cells detach). The cells are recovered in 0.5-1.0 ml of medium and transferred to a 60 x 15 mm (21 cm^2) Falcon Petri dish or directly to a 250-ml (75 cm^2) flask with a screw cap. A small aliquot of this cell suspension can be placed on a new coverslip and incubated 1 hr before medium is added. This coverslip can be used the next day for further chromosomal analysis. When shipment of cells in culture is planned, the cells are planted in 30-ml (25 cm^2) flasks and the flask is filled completely with medium. Shipment is best done at ambient temperature.

C. Cell Morphology

Amniotic fluid cells in culture are known to give rise to at least two morphologically distinct cell populations. One type, which is epithelial-like, proliferates in islands usually for a period limited to two to five passages. These cells adhere strongly to the culture flask and demonstrate increased resistance to trypsin and mechanical breakage. The second type, which can be maintained in culture for over 30 passages has the classical configuration of the fibroblast. Both epithelial and fibroblast-like clones proliferate with equal efficiency in short-term culture. The ratio of one cell type to the other is unpredictable and variable even in cultures derived from the same original sample. Recent work in this laboratory has indicated that morphologically distinct populations of amniotic fluid cells show biochemical differences (67). The need for establishing the normal biochemical properties for each cell type in cultivated amniotic fluid cells is desirable each time a new biochemical parameter is being studied.

D. Preparation of Cells for Enzymatic Assays

Cultivated amniotic fluid cells are rinsed with saline, collected by trypsinization and/or scraping of the flasks, washed three to four times with isotonic saline, and suspended in a small volume of the appropriate buffer. Disruption of the cells is accomplished by freeze-thawing with glass beads three times in a dry ice-acetone bath or nitrogen cavitation, a method in which the cells are exposed to a nitrogen pressure of 800 psi for 20 min followed by a rapid release of pressure until atmospheric pressure is reached. The crude homogenate is centrifuged at 200 x g for 10 min to remove unbroken cells and the supernatant used for the enzyme assays.

In studies utilizing cellular fractions, the cells are suspended in
0.25 M sucrose and disrupted by one of the above methods, and the
homogenate is fractionated by differential centrifugation (68) in a Model
L-Beckman preparative ultracentrifuge. The procedure is as follows:
the homogenate is centrifuged at 600 x g for 10 min to obtain the nuclear
pellet. The supernatant is spun at 4200 x g for 8 min to obtain the mitochondrial pellet. This supernatant is then centrifuged at 30,000 x g for
8 min for the lysosomal pellet. The microsomal pellet is obtained by
centrifuging the supernatant at 60,000 x g for 30 min. The final supernatant
obtained is called the high-speed supernatant.

IV. DETECTION OF CHROMOSOMAL ABNORMALITIES

A. Clinical Commentary

Sex determination and chromosome analysis of amniotic fluid cells
before and after cultivation provides a reliable method to manage pregnancies at risk for cytogenetic aberrations. The indications for study
in our laboratory include chromosomal translocation carriers, maternal
age over 40 years, previous trisomic Down's syndrome, and carriers of
X-linked recessive disorders. In very unusual carrier states, in which
the abnormal chromosomes cannot be identified with certainty or for
which the relation to mental retardation has not been securely established,
karyotyping the amniotic fluid cells will not be helpful at the present time.

B. Staining for Sex Chromatin

Fresh amniotic fluid is spun in a clinical centrifuge at 100 x g for
10 min. The supernatant is discarded and part of the cell button is resuspended in 2 ml of chilled methanol:acetic acid fixative (3:1, v/v) and
incubated at 4°C for 30 min. The suspension is centrifuged at 200 x g for
10 min and the supernatant discarded. The cell pellet is resuspended in
0.25 ml of fresh fixative. One drop of this final solution is mixed with
one drop of 1% cresyl violet (69) or 2% aceto-orcein (61) on a glass microscope slide. The preparation is covered with a coverslip and examined
under a light microscope for the presence of sex chromatin.

C. Fluorescent Staining Technique for the Y Chromosome (70)

1. Reagents

a. Buffer: Dissolve 2.72 g of sodium acetate trihydrous in water,
adjust to pH 5.5 with glacial acetic acid and make up to 100 ml.

b. Quinacrine: Dissolve 500 mg of quinacrine dihydrochloride (Sterling-Winthrop) in 100 ml of water.

c. Fixative: 25% methanol in water.

2. Procedure

A cell pellet is prepared by centrifugation of fresh amniotic fluid at 100 x g for 10 min. The supernatant is discarded and smears are prepared on a number of glass slides. The smears containing 100 to 200 cells are successively:

(a) Fixed in methanol for 15 min.

(b) Stained with quinacrine for 5 min.

(c) Washed in distilled water for 3 min.

(d) Rinsed in buffer.

(e) Mounted in 1:1 glycerol and buffer.

(f) Viewed immediately.

The Y chromosome can be detected as a fluorescent body in interphase nuclei. Between 2 and 10% of the cells may show a fluorescent body if the fetus is a male.

D. Technique of Chromosome Analysis

1. Giemsa Reagents

(a) Giemsa stain; mix 10 g of Giemsa powder in 666 ml of glycerin and 666 ml of methanol.

(b) Sodium phosphate buffer; dissolve 93 mg of $NaH_2PO_4 \cdot H_2O$ and 428 mg of Na_2HPO_4 anhydrous in 300 ml of water.

(c) To 19 ml of sodium phosphate buffer add 1 ml of Giemsa stain and filter twice.

2. Procedure for Amniotic Fluid Fibroblasts

Eighteen hours after a cell-layered coverslip is inverted into a new Petri dish or after subculture onto a new coverslip, 0.1 ml of diacetylmethyl colchicine (Colcemid) (0.01 mg/ml) is added per milliliter of medium. Four to six hours later the medium is removed, the coverslip rinsed gently with hypotonic solution (5 parts distilled water to 1 part fetal calf serum) and then incubated in this hypotonic solution at 37°C for 30 min. The coverslip is gently rinsed in fixative for 20 min. The coverslip is then rinsed in 50% acetic acid and placed cell side up on a slide. A few more drops of 50% acetic acid are added on the preparation which is gently passed back and forth about 5 in. above a flame to promote

spreading of the metaphase plates. The excess acetic acid is drained into
a blotter or tissue. The coverslip is placed in methanol for 5 min and
then stained for 10 min with Giemsa reagent, rinsed with distilled water,
and dried. One or more drops of permount is placed on a clean glass
slide and the stained coverslip, cell-side down, is mounted on this slide.
Between four and six such preparations are mounted for each analysis.
From these preparations, at least 20 metaphases are counted using po-
laroid snapshots. A minimum of two metaphases are then routinely
karyotyped.

E. Procedure for Amniotic Fluid Cells of Epithelial Morphology

When epithelial cells are prepared for chromosome analysis, minor
changes in the fixation process are necessary to allow successful staining.
First, the coverslips are exposed 60 min to the hypotonic solution. Following a quick wash with chilled fixative, the coverslips are incubated in
fresh fixative overnight (18 hr) at 4°C in small Petri dishes. Evaporation
of the fixative is avoided by sealing the dishes with adhesive tape prior
to incubation. The next morning, the coverslips are rinsed once in
methanol and dried. Staining is then performed with Giemsa without
previously spreading the metaphases with acetic acid.

F. Interpretation

The finding of a fluorescent body probably indicates the presence of a
Y chromosome while a sex chromatin body probably indicates the presence of at least two X chromosomes. Failure to detect these bodies does
not necessarily mean that the corresponding chromosome is absent. Uncultivated cells may be damaged and nonviable or the technique itself may
have been inadequate. In our opinion, karyotype analysis of cultivated
amniotic fluid cells is mandatory for accurate determination of fetal sex.

Since the original paper of Steele and Breg (69) reporting successful
karyotype analysis in 4% of their samples, many groups have reported
higher success rates. This rate increased to 90% once the samples were
successfully cultivated. One possible explanation to account for the failure to achieve adequate material for chromosome analysis in cell lines
successfully cultivated is the variation in cell types. Cultivated amniotic
fluid cells of epithelial morphology are completely washed off the cell
preparation when submitted to a fixation which gives optimum spreading
of fibroblast-like cells.

In our laboratory, more than 500 specimens have been cultivated
and chromosome analysis has been accurate in all but three cases. In
each instance, cultures grew more rapidly than usual and chromosome

analysis was performed earlier than usual. In two cases these cells, presumably maternal macrophages, died after approximately one week in culture. Repeat examination of cultures after two weeks will permit accurate chromosome analysis. In cases reported by Uhlendorf (71) maternal cells persisted in culture. In a case reported by Macintyre (72) one of five cultures from one patient had XX cells, whereas the remaining cultures were XY. On the basis of these observations it is suggested that chromosome analysis be performed on at least two cultures at different times of cultivation. The inability to detect the presence of twins of the same or possibly even different sex on the basis of chromosome analysis remains a possible source of error.

Another important problem is the length of time required to obtain sufficient number of cells for adequate cytologic evaluation. This time ranges from 3 to 40 days with an approximate mean of 21 days. Within a single laboratory, the mean time required to complete analysis may be approximately 10 days for a number of months and then increase to 21-25 days. This variation remains unexplained although unproven hypotheses including variations in fetal calf serum, conditions of culture dishes, and variation in lots of medium have been advanced. Methods to reduce the time between procurement of the sample and completion of the analysis are needed to increase the efficiency of this approach for intrauterine diagnosis.

V. PRENATAL DETECTION OF THE GENETIC AMINO ACID DISORDERS

The amino acid disorders in which prenatal diagnosis can be approached through enzyme measurements in cultivated and uncultivated amniotic fluid cells are listed in Table 2. Such an approach has been used in monitoring expectant mother heterozygotes for homocystinuria (73) and maple-syrup-urine disease (74, 75). Cultivated amniotic fluid cells in two of these cases contained normal levels of the correspondent enzyme while the third case showed deficient enzyme activity. The two first children were found to be normal at birth while the third one had maple-syrup-urine disease (75).

The amino acid composition of amniotic fluid has been studied and quantitated both late in pregnancy (76-80) and more recently, early in pregnancy (81-84). Although serial studies were not performed, this information should prove useful for the antenatal diagnosis of certain genetic disorders (Table 3). Frimpter et al. (85) found normal amounts of cystathionine in the amniotic fluid during the last trimester of pregnancy from a woman who previously bore a child with cystathioninuria. The woman subsequently gave birth to a normal child. Morrow et al. (86)

TABLE 2

Hereditary Amino Acid Disorders Detectable in Cultivated (CAF)
and Uncultivated (UCAF) Amniotic Fluid Cells

Disorder	Enzymatic deficiency	Location of enzyme deficiency	Reference
Argininosuccinic aciduria	argininosuccinase	CAF	89
Cystinosis	cystine accumulation	CAF	94, 95
Histidinemia	histidine α–deaminase	CAF	67
Homocystinuria	cystathionine synthetase	CAF	73
Hyperammonemia Type II	ornithine carbamyl transferase	UCAF	90
Hypervalinemia	valine transaminase	CAF & UCAF	91
Ketotic hyperglycinemia	propionyl CoA carboxylase		92
Maple–syrup–urine disease	branched–chain keto–acid decarboxylase	CAF	74, 75
Methylmalonic aciduria	methylmalonyl CoA isomerase	CAF	74, 86
Ornithine– α –keto–acid transaminase deficiency	ornithine– α –keto–acid transaminase	CAF	93

have recently detected methylmalonate in amniotic fluid and maternal urine during the third trimester of pregnancy. The diagnosis of methylmalonic acidemia was confirmed after birth. At 25, 31, and 39 weeks, the methylmalonate concentrations in the amniotic fluid were 0.4, 0.5, and 1.0 mg %, respectively. There was no detectable amount of methylmalonate in control amniotic fluid from normal pregnancies. At 25 and 31 weeks of pregnancy there was respectively 6.7 and 32.6 mg of methylmalonate per day in the mother's urine (normal 5 mg/day). The procedure used in this study consisted of automatic column chromatography (87) of organic acids but gas chromatography (88) may be advantageously substituted for the colorimetric method.

Cystinosis is another disorder in which prenatal diagnosis can be made presently through measurement of the cystine content (94, 95) of cultivated amniotic fluid cells.

VI. PRENATAL DETECTION OF THE LIPIDOSES

A. Gaucher's Disease

1. Commentary

Gaucher's disease is a familial storage disorder which occurs in two main forms: chronic adult and acute infantile. The former is a relatively benign condition characterized by hepatosplenomegaly, hypersplenism, bone and joint manifestations, and poor growth and maturation. The usual onset occurs in middle childhood and the mode of inheritance is autosomal recessive in most instances. In the acute infantile form there is an earlier onset of symptoms with hepatosplenomegaly and neurologic involvement and these infants do not survive beyond the first year of life although occasional patients have lived until middle childhood. The intracellular accumulation of glucocerebrosides in patients with Gaucher's disease is secondary to a deficient glucocerebrose-cleaving enzyme (96, 97). β-Glucosidase has been shown to be deficient in patients with Gaucher's disease. Since this enzyme is present in cultivated skin fibroblasts and amniotic fluid fibroblasts, direct enzyme measurement on cultivated cells provides a method for detecting homozygotes as well as heterozygotes for this disease (98, 99). The assay described here measures the liberation of a fluorescent product (methylumbelliferone) from a synthetic β-glucoside substrate at pH 4.2 (98).

TABLE 3

Free Amino Acids in Human Amniotic Fluid (μmoles/liter)

Amino acid	Weeks of gestation											
	9–12		13–15		16–24		25–32		33–36		36–Term	
	Mean	Range	Mean	Range	Mean	Range	Mean	Range	Mean	Range	Mean	Range
Cysteic acid	6	2–10	9	4–15	8	6–12	7	6–9	10	4–15	11	7–16
Phosphoethanolamine	4	0–19	12	4–19	9	0–31	4	0–14	15	0–19	9	2–17
Taurine	130	58–199	270	72–573	129	43–411	99	65–159	135	88–220	126	43–411
Hydroxyproline	29	3–43	39	30–51	39	19–52	36	22–46	25	11–38	21	9–37
Aspartic acid	9	0–41	35	3–77	11	0–40	3	0–7	3	0–4	2	0–3
Threonine	165	103–230	254	137–333	209	155–276	145	90–191	96	68–156	121	46–194
Serine	54	38–86	47	25–91	30	18–61	37	24–51	27	19–49	75	19–166
Asparagine	15	3–31	24	18–35	26	4–49	6	3–9	5	3–9	4	0–9
Glutamic acid	138	78–213	301	137–614	194	38–536	36	17–56	17	8–23	68	14–327
Glutamine	211	74–417	151	34–293	208	0–380	336	165–578	187	118–399	243	117–379
Proline	158	91–301	139	49–229	156	85–222	160	107–205	103	66–189	116	80–171
Glycine	241	163–388	206	94–387	193	122–271	191	102–290	109	81–169	168	62–418
Alanine	464	266–613	505	325–837	359	248–538	246	156–300	141	92–259	199	104–372

Citrulline	—	—	—	—	12	0–22	—	—	—	—	11	tr–20
α–Amino-n-butyric	—	—	—	—	9	7–11	—	—	—	—	11	tr–19
Valine	212	139–293	262	157–344	130	79–200	63	42–96	43	21–80	56	23–108
Cystine	30	12–69	64	37–85	64	30–155	46	27–70	37	25–51	51	19–123
Methionine	26	14–38	32	18–58	17	8–28	6	2–11	2	0–7	8	tr–19
Cystathionine	—	—	—	—	3	2–4	—	—	—	—	—	—
Isoleucine	52	35–71	64	34–88	29	9–46	9	6–13	6	0–12	14	tr–36
Leucine	134	92–179	142	73–197	65	25–159	20	13–29	13	9–28	31	12–82
Tyrosine	68	52–83	76	47–134	47	25–93	18	12–29	11	7–23	24	6–59
Phenylalanine	83	49–121	82	49–136	56	29–99	21	10–28	12	9–26	27	7–56
β–Alanine	—	—	—	—	—	tr–10	—	—	—	—	—	—
Ethanolamine	16	0–72	57	0–168	30	0–103	26	0–42	40	13–58	39	0–70
Tryptophan	5	0–13	11	0–19	7	0–19	0	0–2	0	0–2	0	0–2
Ornithine	59	36–71	62	37–96	38	18–65	25	15–38	17	12–24	34	9–79
Lysine	338	180–473	332	178–483	204	119–367	118	85–192	88	61–136	119	48–253
Histidine	99	54–141	116	70–169	91	57–128	63	45–111	36	25–71	44	20–80
Arginine	87	59–105	83	39–142	55	28–110	33	20–47	18	14–27	23	tr–56

2. β-Glucosidase Assay

a. Reagents
(a) 0.2 M sodium acetate buffer pH 4.2: Dissolve 2.72 g
$CH_3COONa \cdot 3H_2O$ in water. Adjust the pH to 4.2 with 1 N acetic acid and
make up to 100 ml.
(b) 1.0 mM 4-methylumbelliferyl-β-D-glucopyranoside: Dissolve
8.43 mg in 25 ml of water by boiling for 3 min. Prepare a fresh solution
each time.
(c) 0.25 M sucrose: Dissolve 85.575 g in 1 liter of water and adjust
the pH to 7.0 if necessary.
(d) 0.2 M glycine buffer: Dissolve 15.0 g in water. Adjust to pH
10.7 with 1 N sodium hydroxide in a final volume of 1 liter.

b. Reaction mixture. The reaction mixture contains
(a) 0.02 ml of sodium acetate buffer.
(b) 0.05 ml of 4-methylumbelliferyl-β-D-glucopyranoside.
(c) 0.02 ml of cell preparation containing 0.02-0.04 mg of protein in
0.25 M sucrose or normal saline (e.g., Sec. III, D).

c. Procedure. Duplicate aliquots are incubated for 60 min at 37°C
and the reaction is terminated by the addition of 4.0 ml of 0.2 M glycine
buffer. The 4-methylumbelliferone liberated is measured in an Aminco
Bowman spectrofluorometer at an excitation wavelength of 365 mμ and an
emission wavelength of 450 mμ (Sensitivity setting 18, slit assembly #3)
against an enzyme blank in which glycine buffer has been added without
incubation.

d. Standard. A 1 mM solution of 4-methylumbelliferone is prepared
fresh each time by dissolving 17.6 mg in 100 ml of 0.2 M glycine buffer
from which appropriate dilutions are made to draw a standard curve.

e. Results. Specific activity of β-glucosidase is expressed as
nanomoles 4-methylumbelliferone liberated per hour per milligram of
protein (100).

B. Niemann-Pick Disease

1. Commentary

At least four types of Niemann-Pick disease are presently recognized
on the basis of differences in age and mode of onset, degree of visceral and
neurological involvement, and variations in the pattern of sphingomyelin
and cholesterol accumulation in various organs (101, 102). The use of a

chemically synthesized sphingomyelin-^{14}C has permitted the demonstration of a deficient sphingomyelinase in liver and other tissues from Niemann-Pick patients (103, 104). A study has revealed the enzyme deficiency to be restricted to the infantile (type A) and visceral (type B) forms of the disease (105). Patients with the Nova Scotia variant (type D) show no deficiency of sphingomyelinase activity in their tissues. This observation is in agreement with their slower rate of sphingomyelin accumulation as compared to cholesterol. Cultivated skin fibroblasts (106) and amniotic fluid fibroblasts (102) contain sphingomyelinase activity. The in utero detection of Niemann-Pick disease type A and B has already been made using this approach (107). Holtz et al. (108) has showed that cultivated amniotic fluid cells from Niemann-Pick patients also accumulate an excess sphingomyelin thus suggesting the use of thin-layer chromatography as an alternate to direct enzyme measurement in the prenatal diagnosis of this disease. The assay described here measures the formation of ^{14}C-phosphorylcholine (TCA soluble) from the acid-insoluble ^{14}C-methyl-sphingomyelin substrate (103, 104).

2. Sphingomyelinase Assay

a. Reagents
(a) Cutscum (isooctylphenoxypolyoxyethanol): prepare a 2.5% and a 0.8% solution in water.

(b) 1.0 M sodium acetate buffer pH 5.1: dissolve 13.61 g $CH_3COONa \cdot 3H_2O$ in water, adjust to pH 5.1 with 1 N acetic acid and make up to 100 ml.

(c) 0.4% sodium cholate in water.

(d) 10% human serum albumin in water.

(e) Trichloracetic acid solution, 100% and 20% w/v in water.

(f) 3 mM ^{14}C-methyl-sphingomyelin (400,000 dpm/μmole) in water (109, 110).

(g) Scintillation fluid contains:
    (1) Toluene 2 liters.
    (2) Triton x-100, 1 liter containing 5.5 g
        2,5-diphenyloxazole and 100 mg 1,4-bis
        2'(5-phenyloxazolyl)-benzene.

b. Reaction mixture. The reaction mixture contains:
(a) 0.025 ml sodium acetate buffer.

(b) 0.025 ml sodium cholate.

(c) 0.025 ml 0.8% cutscum.

(d) 0.100 ml cell preparation in 2.5% cutscum: approximately 3.0 x 10^7 cells.

(e) 0.025 ml of ^{14}C-methyl-sphingomyelin (75 mμmoles).

c. Procedure. The incubation is at 37°C for 90 min. The reaction is terminated by the addition of 0.1 ml of human serum albumin, 0.1 ml of 100% TCA, and 0.8 ml of cold water. The tubes are centrifuged at 200 x g for 10 min at 4°C and the supernatant is removed. The precipitates are washed again with 1 ml of cold 20% TCA, centrifuged and the supernatant is added to the first one for liquid scintillation counting in 10 ml of scintillation fluid. The vials are placed in warm water (40°C) until a water-white transparent solution is obtained, kept overnight at 4°C, and counted.

d. Results. One unit of enzyme activity is defined as the amount of enzyme required to catalyze the hydrolysis of 1mμ mole of sphingomyelin per hour per milligram protein (100).

C. Metachromatic Leukodystrophy (MLD)

1. Commentary

This degenerative brain disease begins in infancy and is characterized by ataxia, gait disturbances, and progressive dementia with pyramidal signs, optic atrophy, and convulsions. Death usually occurs by 3 to 4 years of age. Metachromatic material (sulfatide) can be demonstrated in many organs other than brain and may lead to the diagnosis. The biochemical defect has been shown to be a deficiency of a sulfuric acid esterase (111, 112). At least three aryl sulfatases (A, B, and C) can be conveniently detected in tissue by measurement of the hydrolysis of artificial substrates such as p-nitrocatechol sulfate (113). In classic MLD, aryl sulfatase A and cerebroside sulfatase have been found deficient (114). Aryl sulfatase A is diminished in both urine (115) and leukocytes (116) in these patients. In a less common variant, aryl sulfatase A, B, and C are deficient as well as steroid sulfatase (117). Skin fibroblasts from MLD homozygotes and heterozygotes have decreased levels of aryl sulfatase A (118, 119) and this deficiency is present in cultivated amniotic fluid cells (119) permitting prenatal diagnosis (2). The assay described here measures the total sulfatidase activity of a cultivated amniotic fluid cell preparation using a p-nitrocatechol sulfate substrate at pH 5.0.

2. Aryl Sulfatase Assay

a. Reagents
(a) 0.5 M sodium acetate buffer: Dissolve 6.8 g $CH_3COONa \cdot 3H_2O$ in water. Adjust the pH to 5.0 with 1 N acetic acid and make up to 100 ml.
(b) 0.01 M p-nitrocatechol sulfate: Dissolve 31.15 mg in 10 ml of water. Prepare fresh each time.
(c) 0.25 M sucrose: Dissolve 85.575 g in 1 liter of water and adjust to pH 7.0 if necessary.

(d) 2% phosphotungstic acid in 0.1 N HCl.

(e) Alkaline quinol solution; 5 ml of 4% hydroquinone in 0.1 N HCl is added to 100 ml of 2.5 N sodium hydroxide containing 5% sodium sulfite. Prepare fresh each time.

b. Reaction mixture
(a) 0.05 ml of sodium acetate buffer.
(b) 0.10 ml of p-nitrocatechol sulfate.
(c) 0.05 ml of cell preparation in sucrose (e.g., Sec. III, D).

c. Procedure. Three tubes are set for each assay. One tube is acidified with phosphotungstic acid prior to incubation and serves as an enzyme blank. The other tubes are incubated 60 min at 37°C. The reaction is stopped by the addition of 0.75 ml of phosphotungstic acid, and 1.25 ml of alkaline quinol is added as the color developing agent. The p-nitrocatechol liberated is measured at 520 mμ in a Coleman spectrophotometer against the enzyme blank.

d. Standard. A 1 mM stock solution of p-nitrocatechol containing 155 mg/liter of water can be kept refrigerated for up to 6 months without apparent change in its color production. At the time of assay, aliquots of 0.2, 0.1, 0.05, and 0.025 ml of this stock solution are made up to 0.2 ml with water and processed like the reaction mixture.

e. Results. Specific activity is expressed as nanomoles of p-nitrocatechol liberated per hour per milligram of protein (100).

D. Fabry's Disease

1. Commentary

Angiokeratoma corporis diffusum universale (Fabry's disease) is an X-linked recessive disease characterized by onset of limb pain, fever, dysestaesias, abdominal pain, renal and cutaneous abnormalities in middle childhood. Small, keratinized angiomatous skin lesions appear regularly in males on the scrotum, about the umbilicus, and on extension surfaces. Renal involvement includes intermittent proteinuria and hematuria, progressively leading to renal failure, hypertension, and cerebral vascular disease in middle life. All neurons of the autonomic nervous system show extensive accumulation of intracellular lipids. The metabolic defect responsible for the accumulation of this sphingolipid in the tissues, physiological fluids, and culture skin fibroblasts of these patients is due to a deficiency of an enzyme that catalyzes the hydrolysis of the terminal galactose molecule of ceramide trihexoside (120). An elevated acid mucopolysaccharide has also been found in skin fibroblasts in this disorder (121). Leukocytes from homozygotes also show a decreased α-galactosidase activity (122). In plasma from Fabry's patients, Sweely has indicated a deficiency of ceramide trihexoside hydrolysis (123). The presence of α-galactosidase in cultivated human amniotic fluid cells has permitted prenatal diagnosis of this condition (124).

2. α-Galactosidase Assay

a. Reagents
(a) 0.5 M sodium acetate buffer pH 4.5: Dissolve 68 g in water, adjust the pH to 4.5 with 1 N acetic acid and make up to 1000 ml. From this stock solution, prepare a 0.25 M and 0.01 M buffer by diluting 1:1 and 1:50, respectively, with water.
(b) 1 M glycine buffer pH 10.5: Dissolve 7.51 g of glycine in water and adjust the pH to 10.5 with 1 N NaOH. Make up to 100 ml.
(c) 0.01 M 4-methylumbelliferyl-α-D-galactopyranoside: Dissolve 16.9 mg in 5 ml of water.

b. Reaction mixture
(a) 0.050 ml 4-methylumbelliferyl-α-D-galactopyranoside.
(b) 0.025 ml of cell preparation equivalent to 0.2-1.0 mg protein in 0.01 M sodium acetate buffer pH 4.5.
(c) 0.025 ml 0.5 M sodium acetate buffer.

c. Procedure. The incubation is performed at 37°C for 4 hr. The reaction is stopped by the addition of 2.9 ml of glycine buffer and the fluorescence of the 4-methylumbelliferone liberated is measured in an Aminco Bowman Spectrofluorometer against a blank containing water instead of cell preparation. The excitation wavelength is 365 mμ and the emission is at 450 mμ.

d. Standard. A 1 mM solution of 4-methylumbelliferone contains 176.17 mg/liter 0.25 M sodium acetate buffer. From this stock solution a one in ten dilution is made in the same buffer and aliquots of 0.01 ml (1 nmole) to 0.10 ml (10 nmoles) are taken and made up to 3 ml with glycine buffer to serve as standards.

e. Results. Specific activity of α-galactosidase is expressed as nanomoles 4-methylumbelliferone produced per hour per mg protein (100).

E. Globoid Leukodystrophy (Krabbe's Disease)

Krabbe's disease begins in the first months of life with apathy, fretfulness, and gastrointestinal problems. This is followed by seizures, rigidity, dysphagia, and blindness and death occurs within two years after the onset of the symptoms. The basic defect in Krabbe's disease has not been securely established at the present time. Austin (125) reported an increased ratio of cerebroside to sulfatide in the brain of these patients. He also found a deficiency of cerebroside sulfating enzyme (126). More recently a deficiency of galactocerebroside β-galactosidase has been reported in brain and leukocytes from patients with globoid leukodystrophy (128, 129). This enzyme is present in cultivated skin fibroblasts and cultivated amniotic fluid cells. The prenatal diagnosis of Krabbe's disease can now be made on the basis of a deficiency of galactocerebrosidase in cultivated amniotic fluid cells (130).

F. Tay-Sachs Disease (G_{M2} Gangliosidosis)

1. Commentary

This heredodegenerative disorder is the most common of the cerebro-macular degenerative syndromes. It is a disease of infancy with progressive visual loss, spasticity, wasting, arrest of development, seizures, dementia, and early death. It has an autosomal recessive mode of inheritance and occurs most commonly in the Eastern European Jewish population. The gene frequency corresponds to 158 affected newborns per million births among Ashkenazi Jews and 1.7 per million among non-Jews (131). About one in 30 Ashkenazi Jews, and one in 380 non-Jews have been estimated to be heterozygous for the gene of Tay-Sachs disease (132). The G_{M2} ganglioside which accumulates in the gray matter of these patients results from a block in the degradation of this sphingolipid due to an absence of a hexosaminidase. Hexosaminidase A is present in plasma, serum, leukocytes, and skin fibroblasts of normal individuals (133-135). The neonatal diagnosis of Tay-Sachs disease has been made on fibroblasts grown from a biopsy of umbilical cord through the demonstration of an increased ganglioside accumulation (136). Hexosaminidase A is also present in amniotic fluid, uncultivated amniotic fluid cells, and cultivated amniotic fluid cells taken early in pregnancy; the prenatal diagnosis of Tay-Sachs disease has been made on this basis (137).

Sandhoff's disease is clinically indistinguishable from the classic form of Tay-Sachs disease and has been reported in non-Jews. It is also characterized by an excess storage of the G_{M2} ganglioside and its asialo derivative in the central nervous system and the presence of kidney globoside in visceral organs (138, 139). Both hexosaminidase A and B are deficient in various tissues and cultivated skin fibroblasts of patients with this disease (138, 139). Accurate diagnosis of Tay-Sachs disease and Sandhoff's disease can be attained by direct measurement of both hexosaminidase A and B in amniotic fluid, amniotic fluid cells, and cultivated amniotic fluid cells from high risk pregnancies. Electrophoresis on cellulose acetate (140) or acrylamide gel (141) provides a useful tool for diagnosis but in our opinion should be performed conjointly with the enzyme measurement on cultivated amniotic fluid cells (66). The analysis of the ganglioside pattern of cultivated amniotic fluid cells is another alternative which has proven useful in the prenatal diagnosis of Tay–Sachs disease (136).

2. Hexosaminidase A and B Assay

Hexosaminidase A is heat labile at 50°C, whereas hexosaminidase B is stable at 50°C. The amount of each isozyme is therefore determined by assaying the activity with (B) and without (A + B) heat inactivation for 3–4 hr in a water bath (142).

a. Reagents
(a) 0.04 M citrate–phosphate buffer pH 4.4: Dissolve 1.54 g of citric acid (anhydrous) and 1.39 g of dibasic potassium phosphate in water, adjust to pH 4.4 if necessary and make up to 200 ml.

(b) 0.17 M glycine–carbonate buffer pH 9.9: Dissolve 6.33 g of glycine and 9.01 g of sodium carbonate in water, adjust to pH 9.9 and make up to 500 ml.

(c) 1.0 mM 4-methylumbelliferyl-N-acetyl-β-D-glucosaminide: Dissolve 3.95 mg in 10 ml of 0.04 M citrate–phosphate buffer. Substrate must be prepared fresh each time.

b. Reaction mixture
(a) 0.10 ml of 4-methylumbelliferyl-N-acetyl-β-D-glucosaminide.

(b) 0.05 ml of the cell extract containing 0.01 to 0.05 mg protein in citrate–phosphate buffer.

c. Procedure. The incubation is performed at 37°C for 1 hr and terminated by the addition of 5.0 ml of glycine–carbonate buffer pH 9.9. The solution is then read in an Aminco Bowman spectrofluorometer at an excitation wavelength of 365 mμ and emission of 450 mμ against an enzyme blank which has been neutralized prior to incubation.

d. Standard. A 1 mM stock solution of 4-methylumbelliferone is prepared by dissolving 17.6 mg in 100 ml of glycine–carbonate buffer and diluted 100 times with the same buffer. Aliquots of 0.1 to 1.0 ml (1 to 10 nmoles) are made up to 5.15 ml with glycine–carbonate buffer. A standard blank is prepared containing 5.15 ml of the buffer alone.

e. Results. Specific activity is expressed as nanomoles of 4-methyl-lumbelliferone liberated per minute per milligram of protein (100).

3. Electrophoresis of Hexosaminidase A and B (140)

a. Reagents
(a) 0.025 M sodium citrate buffer pH 5.5: Dissolve 7.35 g of sodium citrate in water. Add 82 ml of 0.1 M citric acid solution (19.2 g/liter) and make up to 1 liter.
(b) Staining buffer: Dissolve 9.6 g of citric acid and 13.4 g of dibasic sodium phosphate in water, adjust the pH to 4.5, and make up to 1 liter.
(c) 0.25 M glycine-sodium carbonate buffer, pH 10.0: Dissolve 18.8 g of glycine and 18.8 g of sodium carbonate in water, adjust the pH to 10.0, and make up to 1 liter.
(d) 0.01% w/v 4-methylumbelliferyl-N-acetyl-β-D-glucosaminide in staining buffer.

b. Procedure. Electrophoresis of hexosaminidase is performed using cellulose acetate gel sheets, 17 x 17 cm ("cellogel," Chemetron, Milan, Italy) with a 0.025 M sodium citrate buffer, pH 5.5 in a Shandon U77 electrophoretic tank with filter paper wicks. Ten microliters of the sample are applied slowly on the gel to permit optimal absorption. Electrophoresis is carried out at about 11 V/cm, 10 mA (constant amperage) for two and one half hr at room temperature.

After electrophoresis, the gel sheets are dipped in a solution of 4-methylumbelliferyl-N-acetyl-β-D-glucosaminide in staining buffer. Excess reagent is removed by blotting with filter paper, and the gels are incubated at 37°C for 45 min in a moist chamber. The gels are then sprayed with 0.25 M glycine-sodium carbonate buffer. The fluorescent bands are observed under long-wave UV light. Hexosaminidase A is the fastest moving band and disappears after heat inactivation of the enzyme preparation.

G. Generalized Gangliosidosis (G_{M1} Gangliosidosis)

1. Commentary

This familial neurovisceral lipidosis has its onset early in infancy and presents a wide spectrum of clinical expressions including neurologic and skeletal involvement, gross facial features, and visceromegaly (143). In this autosomal recessive disease there is a neuronal accumulation of G_{M1} ganglioside and ceramide tetrahexoside. The enlarged viscera have been found to accumulate a water-soluble keratan-sulfate type of mucopolysaccharide and a related less soluble sialo mucopolysaccharide (144). The β-galactosidase activity of liver, spleen, kidney, leukocytes, and cultured skin fibroblasts, has been found deficient in these patients (145-149). This enzyme is present in cultivated amniotic fluid cells and its measurement permits the prenatal diagnosis of generalized gangliosidosis. The assay described here utilizes a fluorescent substrate (methylumbelliferyl-galactose) which provides optimal sensitivity although lacking the time saving convenience of the histochemical technique. Starch-gel electrophoresis of the β-galactosidase components of cultivated amniotic fluid cells shows a near absence of both fast and slow moving bands and can provide additional support for the diagnosis (150).

2. β-Galactosidase Assay (151)

a. Reagents
(a) 0.001 M 4-methylumbelliferyl-β-D-galactopyranoside: Dissolve 16.9 mg in 50 ml of water.
(b) 0.5 M sodium acetate buffer pH 5.0: Dissolve 6.8 g CH_3COONa · $3H_2O$ in water, adjust the pH to 5.0, and make up to 100 ml.
(c) 0.2 M glycine buffer, pH 10.7: Dissolve 1.5 g in water, adjust pH to 10.7 with sodium hydroxide and make up to 100 ml.

b. Reaction mixture
(a) 0.10 ml of 4-methylumbelliferyl-β-D-galactopyranoside.
(b) 0.025 ml of sodium acetate buffer.
(c) 0.025 ml of cell preparation (Sec. III, D).

c. Procedure. The incubation is at 37°C for 1 hr and terminated by the addition of 3.0 ml of 0.2 M glycine buffer. An enzyme blank is also run in which glycine buffer is added without previous incubation. The 4-methylumbelliferone formed is read on an Aminco Bowman spectrofluorometer at an excitation wavelength of 365 mμ and an emission wavelength of 450 mμ.

d. Standard. A 1 mM solution of 4-methylumbelliferone contains 17.62 mg/100 ml of sodium acetate buffer. From this solution, a one in ten dilution is made in 0.5 M sodium acetate buffer and aliquots of 0.01 ml (1 nmole) to 0.1 ml are taken and made up to 3.15 ml with glycine buffer. A standard blank is prepared which contains 0.15 ml of sodium acetate buffer and 3.0 ml of glycine buffer.

e. Results. Specific activity of β-galactosidase is expressed as nanomoles of 4-methylumbelliferone formed per hour per milligram protein (100).

3. β -Galactosidase--Histochemical Technique (152)

a. Reagents

(a) Fixative: 25 ml of 50% glutaraldehyde
          0.2 g of potassium chloride
          0.2 g of monobasic potassium phosphate
          1.15 g of dibasic potassium phosphate
          8.5 g of sodium chloride
Dissolve in water, adjust the pH to 7.4, and make up to 1 liter.

(b) Gum-sucrose solution: 1.0 g of gum acacia
                    30.0 g of sucrose
                    10 mg thymol
Make up to 100 ml in water.

(c) Incubation mixture: Dissolve 2 mg of 5-bromo-4-chloroindol-3-yl-β-D-galactopyranoside in 0.5 ml of dimethyl formamide and add 31 ml of 0.1 M sodium acetate buffer pH 5.4 (add 5.6 ml of 0.1 M sodium acetate [1.36 g CH_3COONa·$3H_2O$ to 100 ml in water] to 34.4 ml of 0.1 M acetic acid [1.15 ml of glacial acetic acid to 200 ml in water]), 0.5 ml of isotonic sodium chloride, 8 mg of spermidine hypochloride, and 3.0 ml of 0.05 M potassium ferriferrocyanide (0.823 g ferricyanide and 1.055 g of ferrocyanide to 50 ml in water).

(d) 1% eosin in ethanol.

b. Procedure. The coverslips on which cells have been previously grown are washed three times with saline, dried for 10 min at 37°C and processed as follows:

(a) Dip in the fixative for 20 min at 4°C.
(b) Transfer to cold gum-sucrose solution for 5 min.
(c) Transfer to the incubation mixture for 18 hr or more at 37°C.
(d) Wash in distilled water twice.
(e) Dip 20 sec in 1% eosin.
(f) Rinse again in water and air dry.
(g) Mount in glycerol-gelatin and view under a light microscope.

c. Results. The sites of β -galactosidase activity appear as dark blue granules in the cytoplasm of enzymatically active cells against a pink background.

H. Refsum's Disease

1. Commentary

Heredopathia atactica polyneuritiformis (HAP) is an autosomal recessive disease characterized by cerebellar ataxia, polyneuropathy, retinitis pigmentosa, and cardiovascular, cutaneous, and skeletal alterations. These patients have a deficiency of phytanic acid α-hydroxylase leading to an accumulation of phytanic acid in blood and tissues (153). The antenatal diagnosis of Refsum's disease has not been made presently, but could be approached through the search for an increased phytanic acid concentration in amniotic fluid and the measurement of the rate of oxidation of phytanic acid by amniotic fluid fibroblasts in culture. Since homozygotes for the disease can be separated from heterozygotes and normals with this last method, it should probably be the method of choice.

2. Measurement of ^{14}C-Phytanic Acid Oxidation (154)

Cultures are initiated and maintained in Ham's F-10 medium in which fetal calf serum concentration is maintained at 10% during the preincubation period. The incubation is carried out 48 hr at 37°C in 32-oz bottles containing replicate aliquots of 1×10^7 to 2×10^7 cells in a stationary growth phase, 20 ml of medium, and 0.5 ml of fetal calf serum containing the labeled phytanic acid (0.040 μmoles; specific activity 20-25 μCi/μm) which have been previously filtered through millipore filters (155, 156). This last solution should be clear, indicating completeness of complex formation with albumin. At the end of the incubation, ^{14}CO$_2$ is released from the medium by acidifying to pH 2 with 2 N HCl. The gas in the bottle is drawn slowly through 10 ml of 5.0 N NaOH while being replaced by saline, entering through a second needle. The NaOH containing Na$_2$^{14}CO$_3$ is slurried with 3 g of anthracene crystals (fluorescense grade, x 480, Eastman Kodak Company) and 0.2 ml of a solution of Triton, Gr-5, diluted 1:10 v/v (Rohm & Haas Co.), and counted in a liquid scintillation spectrometer.

Results. Oxidation of phytanic acid is expressed as the percentage of phytanate-U-^{14}C oxidized per 10^7 cells per 48 hr. The observed rate of phytanate oxidation in cultures derived from HAP patients is about 3% of that in control cultures. Heterozygotes for the gene show about 60% of the normal values.

VII. PRENATAL DETECTION OF THE MUCOPOLYSACCHARIDOSES

A. Commentary

Fratantoni et al. (157) have reported the prenatal detection of Hurler's and Hunter's diseases on the basis of staining and metabolic studies of cells cultured from amniotic fluids. Matalon has been able to diagnose Hurler's syndrome in utero by demonstrating excessive amounts of dermatan sulfate and the presence of heparitin sulfate in the amniotic fluid

(158). The level of acid mucopolysaccharides in amniotic fluid is known to decrease with time in normal pregnancies. In an amniotic fluid obtained at the 14th week of a pregnancy which resulted in the delivery of a baby with Hurler's syndrome, 0.087 mg of mucopolysaccharides per ml of amniotic fluid was present compared to a normal of approximately 0.02 mg/ml.

There are presently three approaches to the prenatal detection of the mucopolysaccharidoses: (1) measurement of the level of acid mucopoly-saccharides in amniotic fluid, (2) kinetic studies with $^{35}SO_4$, and (3) demonstration of metachromasia in cultivated amniotic fluid cells. The methodology for column fractionation and colorimetric determination of acid mucopolysaccharides by the Carbazole reaction is given elsewhere in this book. This procedure can be applied directly to specimens of amniotic fluid from which the cells have been removed by centrifugation.

B. Staining for Acid Mucopolysaccharides

The initial studies on fibroblasts cultured from affected individuals and heterozygous carriers of the genetic mucopolysaccharidoses showed that the demonstration of metachromasia was correlated with an increased intracellular concentration of glycosaminoglycans. It must be remembered that the dye toluidine blue (159) reacts nonspecifically with negatively charged substances including lipids, nucleic acids, polypeptides, and meta-phosphates (160). Table 4 lists the various genetic disorders which may give a positive reaction to toluidine blue in tissue culture. For this reason, it is evident that methachromasia by itself has no place presently in the prenatal detection of the genetic mucopolysaccharidoses (157). However, it remains a useful tool when used in correlation with more reliable biochemical techniques. Staining with Alcian blue (163) has also been found to be a useful supplement to metachromasia to detect increased intra-cellular concentrations of mucopolysaccharides.

1. Toluidine Blue Staining Procedure (157)

a. Reagents
(a) 0.1% toluidine blue O (Matheson, Coleman and Bell, color index 52040) in 30% methanol in deionized water (filtered twice).
(b) Acetone-xylene mixture (1:1).
(c) Normal saline.

b. Procedure
Cells are grown on coverslips in small Petri dishes for 48–72 hr. The medium is discarded and the coverslips are removed from the dish and processed immediately:

TABLE 4

Genetic Disorders in Which Metachromasia
Can Be Demonstrated in Skin Fibroblast Cultures (161, 162)

Content in Mucopolysaccharides	
Normal	Elevated
Morquio's syndrome	Hurler's syndrome
Maroteaux-Lamy	Hunter's syndrome
Cystic fibrosis	Sanfilippo syndrome
Pseudoxanthoma elasticum	Scheie's syndrome
Gaucher's disease	Marfan's syndrome
Myotonic muscular dystrophy	Generalized gangliosidosis
Multiple epiphyseal dysplasias	Larsen's syndrome
Carotinemia	α-Fucosidase deficiency
Jakob-Creutzfeldt disease	Krabbe's disease
Epidermolysis bullosa	Fabry's disease
Scleroderma	I-Cell disease
Vitamin D deficient rickets	Late infantile amaurotic idiocy
Normals (6-27%)	α-Mannosidase deficiency
	Lipomucopolysaccharidosis

(a) Rinse twice in warm saline (5 sec).
(b) Dip in toluidine blue for 4 to 5 min.
(c) Rinse in acetone twice.
(d) Rinse in acetone-xylene.
(e) Rinse in xylene.
(f) Mount immediately.

c. Results. Metachromatic material appears as intracellular reticulum, granules, or large inclusions, red or purplish-red in color.

2. Alcian Blue Staining Procedure (163)

a. Reagents
(a) 1% Alcian blue 8GX (Imperial Chemical Industries, Ltd., Teesside,
England) in distilled water.

(b) 4 M magnesium chloride (81.3 g $MgCl_2 \cdot 6H_2O$ and add water to make up to a 100 ml volume.

(c) 1 M sodium acetate, pH 5.7 (13.6 g $CH_3COONa \cdot 3H_2O$ in water). Adjust to pH 5.7 and make up to 100 ml.

(d) Normal saline.

(e) 70% and 100% ethanol.

(f) Xylene.

b. Procedure. The coverslips carrying the cells are washed quickly in warm saline and immersed at room temperature overnight in an upright Columbia coplin jar (A. H. Thomas Co., Philadelphia) containing 10 ml of dye solution containing:

(a) 0.75 ml $MgCl_2$.

(b) 0.25 ml sodium acetate.

(c) 0.50 ml Alcian blue.

(d) 8.5 ml of water.

The next morning the coverslips are dehydrated in 70% and 100% ethanol, cleared in xylene, and mounted.

c. Results. Alcianophilia (blue color) can be detected in the cytoplasm of fibroblast cultures derived from patients with the following diseases: Hurler's, Hunter's, Morquio's, Scheie's, Maroteaux-Lamy, generalized gangliosidosis, and Larsen's syndrome. There is no alcianophilia in either Sanfilippo or Marfan's syndromes even if the acid glycosaminoglycan content is elevated in cultivated fibroblasts from these patients.

C. Radioactive Sulfate Incorporation (164)

This simple assay measures the incorporation of $^{35}SO_4$ into mucopolysaccharides since mammalian cells generally do not incorporate sulfate into methionine and cystine residues of protein (165).

a. Reagents

(a) $H_2S^{35}O_4$ (New England Nuclear Corp.).

(b) 80% ethanol.

(c) 10% sodium hydroxide.

(d) Scintillation fluid made of 4 g of 2,5-diphenyloxazole, 0.05 g of 1,4-bis-2' (5-phenyloxazolyl)-benzene in 1 liter of toluene-methylcellosolve, 1:1, v/v.

b. Procedure. Replicate aliquots (one million) of cultivated amniotic fluid cells are planted in two 100 mm Falcon plastic Petri dishes and incubated in Eagle's minimal essential medium (in which magnesium chloride is substituted for magnesium sulfate) containing 3-15 x 10^6 cpm of carrier-free $H_2S^{35}O_4$ per ml, 15% fetal calf serum, and antibiotics. After two and four days of incubation, one of the dishes is processed:

    (a) Remove the medium.

    (b) Wash the cells twice in Puck's saline.

    (c) Trypsinize for 15 min at 37°C.

    (d) Recover the cells in Puck's saline.

    (e) Centrifuge at 200 x g for 15 min.

    (f) Discard the supernatant.

    (g) Extract 4 times with 2 ml of boiling 80% ethanol for 1 min and discard the extracts.

    (h) Dissolve the residue by heating in 1.2 ml of 10% NaOH.

    (i) Measure the proteins on 0.1-ml aliquots (100).

    (j) Determine the radioactivity on 0.5-ml aliquots in 10 ml of scintillation fluid.

    c. Results. The cpm of ^{35}S accumulated per microgram of cell protein are plotted as a function of time. Cells from affected individuals accumulate $^{35}SO_4$-labeled mucopolysaccharides at a linear rate, whereas normal cells reach a steady state within two days. Cells from heterozygotes display an essentially normal pattern.

D. Autoradiography with Radioactive Sulfate (166)

Cells derived from susceptible and control individuals are grown directly on glass coverslips in the same medium used for the kinetic studies described previously. For autoradiography, the coverslips are incubated for 24 hr in 3 ml of medium containing 10 μCi of carrier-free $Na_2^{35}SO_4$ (New England Nuclear Corp.). After removal of the medium, the coverslips are washed with Puck's saline and successively:

    (a) Fixed with methanol for 5 min.

    (b) 0.5% cetyltrimethyl-ammonium bromide in 10% formalin for 3 hr.

    (c) Covered with Kodak AR 10 films or NTB 3 emulsion at 4°C for 2-4 weeks, developed and stained with haematoxylin or Giemsa.

In the control cultures, autoradiographic granules are few and scanty so that it is hard to differentiate them from the background grains. In the cultures from Hurler's and Morquio's diseases there is a characteristic presence of grains exclusively located in the cytoplasm. In neither of these two diseases is there enough data available on cultivated amniotic fluid cells to permit a satisfactory delineation between carriers and homozygotes.

VIII. PRENATAL DETECTION OF CARBOHYDRATE DISORDERS

A. Galactosemia

1. Commentary

In this disorder, transmitted as an autosomal recessive trait, there is impaired ability to metabolize galactose. This is caused by the absence of the enzyme phosphogalactose uridyl transferase. The enzyme is present in cultivated amniotic fluid cells and provides a method for the diagnosis of galactosemia in utero (2). In children whose exposure to galactose has been adequately controlled from before birth, it seems likely that the prognosis is good. Such prenatal control of galactose uptake in the mother can prevent neonatal cataracts, permit normal liver function and development (167). The in utero detection of galactosemia has been established utilizing the quantitative assay of transferase activity in cultivated amniotic fluid cells.

2. Galactose-1-Phosphate Uridyl Transferase Assay (168)

a. Reagents
(a) 19.5 mM dithiothreitol (Cleland's reagent), dissolve 3 mg/ml of water.

(b) 0.5 M glycylglycine buffer pH 8.7; 6.6 g of glycylglycine are dissolved in water, the pH adjusted to 8.7 with 1 N NaOH, and made up to a final volume of 100 ml.

(c) 0.01 M glycylglycine buffer pH 7.5; 1.32 g are dissolved in a final volume of 1 liter of water after pH adjustment.

(d) Magnesium chloride; dissolve 32 mg of $MgCl_2 \cdot 6H_2O$ in 100 ml of water.

(e) 0.5% nicotinamide adenine dinucleotide phosphate (NADP); 5 mg/ml of water.

(f) 6 mM uridine-5'-diphosphoglucose (UDPG); 4.12 mg/ml of water.

(g) 2 mM galactose-1-phosphate (Gal-1-P); 8.7 mg/ml of water.

(h) Auxiliary enzymes; Phosphoglucomutase 2 μl, glucose-6-phosphate dehydrogenase 20 μl, and 6-phosphogluconic dehydrogenase 2 μl in 5 ml of water.

b. Cell Preparation
The cells are harvested by trypsinization, washed twice with cold normal saline and disrupted by freezing and thawing 4 times in a small

volume of 0.01 M glycylglycine buffer pH 7.5, and centrifuged at 30,000 rpm for 30 min at 4 °C and the supernatant is used for assay and protein determination (100).

c. Reaction mixture. Contains
(a) 0.1 ml of 0.5 M glycylglycine buffer pH 8.7.
(b) 0.1 ml of dithiotheitol.
(c) 0.1 ml of cell preparation.
Incubate at 37°C for 30 min. Add 0.7 ml of a mixture containing:
0.1 ml of NADP.
0.1 ml of MgCl$_2$.
0.1 ml of Gal-1-P.
0.1 ml of auxiliary enzymes.
0.1 ml of UDPG.
0.2 ml of water.
Except in the blank tube which receives 0.1 ml of water in place of Gal-1-P. The rate of NADP reduction is recorded at 340 mμ in a Gilford automatic spectrophotometer. The specific activity is expressed as μ moles of UDPG consumed per hour per mg of protein.

B. Type II Glycogenosis (Pompe's Disease)

1. Clinical Commentary

Pompe's disease is a fatal autosomal recessive disorder characterized by the accumulation of glycogen in heart, skeletal muscle, brain, liver, and kidney (169). Symptoms related to impaired cardiac function may be manifest shortly after birth but in most instances become evident within the first six months of life. The infiltration of glycogen in other tissues may produce clinical manifestations which mimic diseases like Down's syndrome, cretinism, and amyotonia congenita on the basis of extreme muscular hypotonicity and thickening of the tongue. Massive infiltration of the nervous system may suggest vascular or degenerative disorders of the brain.

Hers (170) initially demonstrated the deficiency of lysosomal acid maltase, α-1,4 glucosidase in the liver of patients with Pompe's disease. The disease has been detected in utero on the basis of a deficiency of α-1,4 glucosidase in cultivated and uncultivated amniotic fluid cells (171-173). In one of these cases (171) no α-1,4 glucosidase was present in the amniotic fluid while in the other case normal levels were detected (173).

Recent studies in this laboratory (174) suggest that the in utero diagnosis of Pompe's disease must rest on the presence or absence of α-1,4-glucosidase activity in cultivated amniotic fluid cells and not on amniotic fluid alone. The enzyme in amniotic fluid differs from that found in other tissues in several properties including pH optimum and inhibition by turanose.

2. α-Glucosidase Assay

α-1, 4-Glucosidase is assayed by the method of Nitowsky and Grunfeld (175) by measuring the glucose liberated from maltose by the glucose oxidase method of Dahlquist (176).

a. Reagents
(a) 0.1 M sodium acetate buffer pH 4.0:
    (1) Dissolve 1.36 g of sodium acetate trihydrous in 100 ml of water.
    (2) Prepare a 0.1 M acetic acid solution by diluting 1.15 ml of glacial acetic acid to 200 ml with water.
    (3) Mix 9 ml of (1) and 41 ml of (2) and adjust pH to 4.0 if necessary.
(b) 1% Maltose: Dissolve 1 g in 100 ml of 0.1 M sodium acetate buffer pH 4.0.
(c) 0.5 M tris buffer pH 7.0: Dissolve 61 g of tris in 85 ml of 5 N hydrochloric acid and water to a final volume of 1 liter.
(d) 5 N hydrochloric acid: Mix 86.1 ml of concentrated hydrochloric acid and water to a final volume of 200 ml.
(e) TGO reagent: Prior to using, mix 100 ml of (1) with 0.5 ml of (2):
    (1) 62.5 mg glucose oxidase (Sigma Type II): 2.5 mg peroxidase (Sigma Type II) in 500 ml of 0.5 M tris buffer pH 7.0 (make up fresh weekly).
    (2) 100 mg of O-dianisidine (Sigma) in 10 ml of 95% ethanol.

b. Reaction mixture
Each tube contains
    (1) 0.05 ml of maltose.
    (2) 0.10 ml of water.
    (3) 0.10 ml of cell preparation equivalent to 0.2 mg protein.

c. Procedure. The incubation is performed at 37°C in a water bath. After 1 hr, 1.75 ml of water is added and the tubes are boiled for 2 min, cooled, and centrifuged 10 min at 600 x g in the cold. Aliquots of 0.5 ml of the supernatent are incubated with 3.0 ml of TGO reagent at 37°C for 1 hr. The color produced is read in a Coleman spectrophotometer at 420 mμ.

d. Standards. From A stock 0.2% glucose standard solution containing 100 mg of dextrose in 500 ml of water, the following dilutions are prepared to which 3.0 ml of TGO reagent are added before the final incubation.

Stock glucose solution, ml	Water, ml	μg of glucose, per tube (3.5 ml)
0.00	0.50	0.0
0.05	0.45	10.0
0.10	0.40	20.0
0.20	0.30	40.0
0.30	0.20	60.0
0.40	0.10	80.0

e. Results. The activity is expressed as micromoles of maltose hydrolyzed per minute per gram protein.

For normal values and comparisons of the specific activities of α-1, 4-glucosidase in various fetal tissues and amniotic fluid cells the reader is referred to the recent paper by Salafsky and Nadler (174).

C. Fucosidosis

1. Clinical Commentary

Some children with the clinical appearance of Hurler's syndrome have been described (177) with excessive storage of fucose-containing glycolipids. The lysosomal enzyme α-fucosidase was shown to be deficient in their liver, brain, lung, and kidney (178). This enzyme is present in cultivated skin fibroblasts and cultivated amniotic fluid cells and makes this disorder diagnosable before birth.

2. α-Fucosidase Assay (179)

a. Reagents
(a) 0.5 M sodium acetate buffer pH 5.5: Dissolve 6.8 g of CH$_3$COONa ·3H$_2$O in water. Adjust pH to 5.5 with 1 N acetic acid. Make up to 100 ml with water.

(b) 15 mM p-nitrophenyl-α-L-fucoside (Sigma Chemical Co.): Dissolve 85.4 mg in 20 ml of sodium acetate buffer.

(c) 1 M glycine buffer pH 10.7: Dissolve 7.51 g of glycine in water. Adjust pH to 10.7 with 1 N sodium hydroxide and make up to 100 ml with water.

b. Reaction mixture. Contains
(a) 0.1 ml of cell preparation in 0.25 M sucrose containing an appropriate amount of protein.

(b) 0.2 ml of substrate p-nitrophenyl-α-L-fucoside.

(c) Incubate for 2 hr at 37°C. Stop the reaction by adding 2.7 ml of glycine buffer and read the p-nitrophenolate ions liberated in a spectrophotometer at 410 mμ against an enzyme blank in which glycine has been added prior to incubation.

c. Standard. A 1 mM solution of p-nitrophenol (6.95 mg in 50 ml of water) is diluted as necessary and serves as a reference.

d. Results. α-Fucosidase activity is expressed as millimicromoles of p-nitrophenol liberated per milligram protein (100) per hour.

D. Farber's Disease

1. Clinical Commentary

Patients with this disorder show signs and symptoms similar to those seen in Hurler's syndrome, but accumulate mannose and glucosamine (180, 181). The lysosomal enzyme α-mannosidase has been found deficient in these patients, while other lysosomal enzymes including α-fucosidase, β-galactosidase, and β-glucuronidase all have abnormally high levels of activity. Since α-mannosidase activity is normally detectable in cultivated amniotic fluid cells, the prenatal diagnosis of this condition is now possible.

2. Technical Commentary (179)

Assay of the lysosomal enzymes α-mannosidase, β-galactosidase, and β-glucuronidase can be performed using the p-nitrophenyl derivatives of these sugars under conditions similar to those described for α-fucosidase. A 15 mM substrate concentration is above the saturation limit of these enzymes and can be used for all assays. The pH optimum of these enzymes is broadly distributed between 5.0 and 6.5. Therefore, in order to facilitate the handling of small samples, the assays can be performed at pH 5.5.

IX. PRENATAL DETECTION OF MISCELLANEOUS DISEASES

A. Lesch-Nyhan Syndrome

1. Commentary

Lesch-Nyhan syndrome is an X-linked recessive familial disorder characterized by spasticity, choreo-athetosis, psychomotor retardation, compulsive automutilation, and hyperuricemia and hyperuricosuria (182). The basis for this disorder is a virtual lack of hypoxanthine-guanine phosphoribosyltransferase (HGPRT) which converts hypoxanthine and guanine to their respective nucleotides (183). A partial deficiency of this enzyme leads to excessive uric acid production and the development of uric acid calculi and gouty arthritis. These latter patients exhibit mild neurologic symptoms in about 20% of the cases (184). McDonald and Kelley have reported a patient with classical features of the Lesch-Nyhan syndrome whose red blood cells exhibited increased HGPRT activity at very high concentration of magnesium 5-phosphoribosyl-1-pyrophosphate (185). The enzyme HGPRT is normally detectable in a number of different human tissues as well as in cultivated skin fibroblasts and amniotic fluid cells (183, 186, 187, 189). Fibroblasts from patients whose erythrocytes show the incomplete deficiency of HGPRT demonstrate a slightly higher activity than that present in the complete Lesch-Nyhan syndrome. Furthermore, the the mutant HGPRT has a greater heat lability than the normal enzyme (188). A number of methods are available for the in utero detection of the Lesch-Nyhan syndrome including autoradiography and assay of the enzyme on cultivated amniotic fluid cells and possibly direct uric acid measurements on amniotic fluid (189-191).

2. Hypoxanthine-Guanine Phosphoribosyltransferase Assay (188)

a. Reagents
(a) 0.5 M Tris buffer pH 7.4: Dissolve 6.06 g of Tris (hydroxymethyl) aminomethane in water, adjust the pH to 7.4, and make up to 100 ml.

(b) 0.1 M magnesium chloride: 2.03 g of $MgCl_2 \cdot 6H_2O$ are dissolved in water to a final volume of 100 ml.

(c) 0.009 M 5-phosphoribosyl-1-pyrophosphate (Calbiochemical) 42.36 mg to 10 ml in water.

(d) 4.5 mM hypoxanthine-8-^{14}C (Schwarz Bio-Research Inc.) 4-5 mCi/mmole: Dissolve 2.72 mg of cold hypoxanthine and 100 μCi of hypoxanthine-8-^{14}C in a final volume of 5 ml in water.

(e) 0.1 M EDTA; 3.72 g of sodium (di) ethylenediamine tetraacetate are dissolved in a final volume of 100 ml of water.

(f) Inosinic acid: 1 mg/ml for marker solution.

(g) Inosine: 1 mg/ml for marker solution.

(h) 0.05 M borate, 0.001 M EDTA buffer, pH 9.0; Dissolve 372 mg of EDTA and 19.1 g of sodium borate in water, adjust to pH 9.0, and make up to 1 liter.

(i) Thymidine triphosphate 0.028 M: Dissolve 14 mg of thymidine-5-triphosphate, trilithium, hexahydrate per ml of water.

b. Reaction mixture
(a) 0.010 ml of 0.5 M Tris buffer, pH 7.4.

(b) 0.005 ml of 0.1 M $MgCl_2$.

(c) 0.010 ml of 5-PRPP.
(d) 0.015 ml of hypoxanthine-8-^{14}C.
(e) 0.010 ml of thymidine triphosphate.
(f) 0.050 ml of cell extract in 0.01 M Tris buffer, pH 7.4.

c. Procedure. The mixture is incubated at 25°C for 60 min in a
shaking water bath and the reaction terminated by the addition of 0.02 ml
of 0.1 M EDTA followed by freezing in dry ice-acetone. After thawing,
0.02-ml aliquots are spotted at 1-in. intervals on a Whatman 3 MM paper
previously spotted with 0.02 ml of inosine and inosinic acid. High voltage
electrophoresis is carried out at 4000 V for 20 to 30 min in borate EDTA
buffer, pH 9.0. The areas corresponding to inosinic acid and inosine are
located under ultraviolet light, cut out, and counted in a liquid scintillation
counter in Bray's solution (192).

d. Results. Specific activity of PRT is expressed as millimicromoles of hypoxanthine-8-^{14}C converted to inosinic acid and inosine per
milligram protein (100) per hour at 25°C.

3. Autoradiography with Hypoxanthine-^{3}H

Plate the cells at low density in small Petri dishes containing a glass
coverslip and add 2-3 ml of supplemented Ham's F-10 medium. After 24
hr of culture, the medium is removed and replaced by 2 ml of medium
containing 20 μCi of hypoxanthine-^{3}H (New England Nuclear, specific
activity 3.14 μCi/mmole). After another 24 hr, the coverslips are removed and dipped successively into small baths containing:
(a) Sodium chloride 0.154 M.
(b) Absolute methanol for 5 min.
(c) Ice cold 5% trichloracetic acid for 25 min.
(d) Distilled water.
and dried at room temperature. The processed coverslips are then coated
with Kodak NTB 3 emulsion and exposed for 3 days in a refrigerator. The
coverslips are then developed and treated with Giemsa stain. The silver
grain density is observed under a light microscope and counted, if necessary. Lesch-Nyhan homozygotes show 10 to 20% the granularity of normal
controls while heterozygotes show normal, intermediate, or low density.

B. Lysosomal Acid Phosphatase Deficiency

1. Commentary

Lysosomal acid phosphatase deficiency (LAPD) is an autosomal recessive disorder characterized by vomiting, lethargy, opisthotonos, terminal

bleeding, and death in early infancy (193). Acid phosphatase activity in fibroblasts from patients with this disorder is decreased to 30% of normal in the original homogenates and to less than 2% in the lysosomal fraction. Heterozygotes can be identified in cultivated fibroblasts as well as lymphocytes stimulated with phytohemagglutinin. In addition, cultivated amniotic fluid cells have been utilized to establish the in utero diagnosis of this disorder (2). Total acid phosphatase deficiency (TAPD) has been observed in a patient dying at 36 hr of life from similar clinical manifestations. Acid phosphatase activity in his fibroblasts was reduced to 2% of normal in the original homogenate and could not be detected in the lysosomal fraction.

2. Acid Phosphatase Assay

Acid phosphatase mediates the phosphorolysis of ortho-phosphate from phosphate esters in acidic buffer systems. The enzyme is assayed by the procedure using p-nitrophenylphosphate or β-glycerophosphate as substrates.

a. p-Nitrophenylphosphate Method (194)

(1) Reagents
      (a) 0.05 M p-nitrophenylphosphate in 0.1 M acetate buffer, pH 4.8: 185 mg of p-nitrophenylphosphate is dissolved in 10 ml of 0.1 M sodium acetate buffer pH 4.8.
      (b) 0.1 M sodium acetate buffer: Dissolve 1.36 g CH_3COONa $\cdot 3H_2O$ in water, adjust pH to 4.8 with 1 N acetic acid, and make up to 100 ml.
      (c) 0.25 M sodium hydroxide: 1 g in 100 ml of water.

(2) Reaction mixture
      (a) 0.2 ml of 0.05 p-nitrophenylphosphate in acetate buffer.
      (b) 0.2 ml of cell preparation in 0.25 M sucrose (equivalent to 0.020-0.150 mg of protein).

(3) Procedure. The incubation is at 37°C for 30 min and the reaction is stopped by the addition of 3.0 ml of 0.25 N NaOH. The p-nitrophenol liberated is measured at 410 mμ in a Coleman spectrophotometer against an enzyme blank in which NaOH has been added without incubation.

(4) Standards. A 1 mM solution containing 6.95 mg p-nitrophenol in 50 ml of water is diluted as necessary and serves as a reference.

(5) Results. Specific activity is expressed as micromoles of p-nitrophenol liberated per milligram protein per hour.

b. β -Glycerophosphate Method (195)

(1) Reagents
(a) 0.05 M sodium acetate buffer: Dissolve 6.8 g $CH_3COONa \cdot 3H_2O$ in water, adjust the pH to 5.0 with 1 N acetic acid, and make up to 1 liter.

(b) 0.02 M sodium β-glycerophosphate pentahydrate: Dissolve 630 mg in 100 ml of 0.05 M sodium acetate buffer pH 5.0.

(c) 10% trichloracetic acid.

(2) Reaction mixture
(a) 0.90 ml of β-glycerophosphate in sodium acetate buffer.

(b) 0.10 ml of cell preparation in sucrose (equivalent to 0.1 mg protein).

(3) Procedure. The incubation is at 37°C for 60 min. The reaction is terminated by the addition of 1.0 ml of 10% ice cold trichloracetic acid. The tubes are centrifuged at 200 x $\underline{g}$ for 10 min and the supernatant is then assayed for inorganic phosphate liberated by a modification of the method of Fiske and Subbarow.

c. Phosphate Determination

(1) Reagents
(a) Ammonium molybdate: Dissolve 5 g of ammonium molybdate in about 50 ml of water. Add 200 ml of 5 N sulfuric acid and make up to 400 ml with water.

(b) Fiske and Subbarow reducer (Sigma Chemical Co.): Dissolve 0.5 g of reducer in 6.3 ml of water and keep in a dark brown bottle.

(c) Standard phosphorus solution (Sigma Chemical Co.) containing 20 μg of phosphorus as KH_2PO_4 per ml.

(2) Procedure. Aliquots of 0.5 and 1.0 ml of the reaction mixture and from 0.1 to 0.5 ml of phosphorus standard are made up to 2.0 ml with water. 0.4 ml of molybdate and 0.2 ml of reducer are added and the tubes shaken gently. The optical density is read at 700 mμ after 10 min.

3. Acid Phosphatase-Histochemical Method (196)

a. Reagents
(a) Fixative: 0.5 g of calcium chloride
                  1.07 g of sodium cacodylate
                  4.0 ml of 50% glutaraldehyde
Dissolve in water, adjust to pH 7.4, and make up to 100 ml.

(b) Incubation medium:
        Solution A: 0.4 g of sodium β-glycerophosphate

6.5 g of sucrose

1.07 g of sodium cacodylate

Dissolve in water, adjust to pH 5.2, and make up to 100 ml.

Solution B: 0.3 g of lead nitrate

6.5 g of sucrose

1.07 g of sodium cacodylate

Dissolve in water, adjust to pH 5.0 and make up to 100 ml.
Add 15 ml of A dropwise to 10 ml of B and let stand (16 hr) overnight at
25°C and 1 hr at 37°C. Filter before use.

(c) Sucrose–cacodylate buffer pH 5.0 and 7.4.

6.5 g of sucrose

1.07 g of sodium cacodylate

Dissolve in water, adjust to pH 5.0 and 7.4, and make up to 100 ml.

(d) 0.4% yellow ammonium sulfide.

b. Procedure. The cells grown on a glass coverslip are washed
three times with normal saline at room temperature and processed successively following the procedure which follows:

(a) Dip 12 min at 4°C in fixative.

(b) Wash six times, for 20 min each, in distilled water.

(c) Let stand in sucrose–cacodylate buffer pH 7.4 overnight at 4°C
and 30 min at 37°C.

(d) Transfer to the incubation mixture at 37°C for 3–4 hr.

(e) Wash briefly with distilled water.

(f) Wash three times for 30 min in sucrose–cacodylate buffer pH
5.0 at 4°C.

(g) Treat with ammonium sulfide for 3 min.

(h) Wash six times for 5 min in water.

(i) Mound in glycerine jelly and view.

c. Results. The sites of acid phosphatase activity will be seen as
black granules of lead nitrate in the cytoplasm.

X. PRENATAL DETECTION OF THE CONGENTIAL NERVOUS
SYSTEM MALFORMATIONS

Major congenital malformations leading to and associated with mental
retardation or early death are, at present, essentially not detectable in
early pregnancy. Many potential approaches to prenatal diagnosis of
severe fetal anomalies are possible through direct and/or indirect visualization of the fetus. These include: (1) roentgenogram, (2) amniography
or fetography, (3) amnioscopy or fetoscopy, and (4) ultrasonic scanning.

A. Roentgenograms

A number of congenital malformations have been detected usually late in the third trimester of pregnancy using this approach (197). Anencephaly, hydrocephalus, microcephaly, encephalocele, meningocele, and many other skeletal anomalies have all been detected using simple radiological techniques. This approach is extremely useful to the obstetrician in determining optimal methods of delivery but has been rarely useful in the antenatal detection of genetic disorders early in pregnancy.

B. Amniography

The technique of amniography or fetography requires amniocentesis with injection of a contrast medium into the amniotic cavity (198, 199). Fetal soft tissue abnormalities displace the opacified amniotic fluid in such a way as to delineate fetal abnormalities such as meningomyelocele. Oil-soluble dyes have been recently introduced (199, 200) into the amniotic cavity with obvious advantages over water-soluble dyes in certain cases. Their increased affinity for vernix caseosa and low diffusion eliminating glooming of the amniotic fluid, suggests that they may permit better delineation of fetal outline.

C. Amnioscopy

Direct visualization of the fetus using a fine endoscope or a fiberoptic instrument has been reported prior to therapeutic abortion and in severely affected Rh-sensitized pregnancies (201, 202). This method referred to as fetoscopy would provide an optimal approach for detection of common congenital malformations if applied earlier in pregnancy.

D. Ultrasonic Scanning

Although ultrasonic scanning has permitted the intrauterine detection of polycystic kidneys (203), this technique has been used primarily for placental localization (204, 205) in the latter half of pregnancy. It could theoretically be used to diagnose alterations in fetal head and body size whenever manifested: e.g., as early as 24 weeks of pregnancy (206).

XI. CONCLUSION

There is an increasing number of disorders leading to mental retardation in which prenatal diagnosis and treatment will be attained in the next few years. However, no one investigator is going to see very many patients with these disorders. Despite this fact, the benefits of amniocentesis and prenatal diagnosis should be provided to all parents at "high risk" for having children with these disorders. Hopefully, every laboratory and hospital will not attempt to duplicate facilities already available in trying to monitor "high risk" pregnancies for rare biochemical disorders, but will send the appropriate material to nearby centers that have gained expertise with the particular disorder. It is also impractical for one laboratory to set up enzyme assays and methodologies for the occasional patient. Someone having knowledge and experience in handling fetal material is the person of choice to perform prenatal genetic diagnosis.

ACKNOWLEDGMENTS

The development of some of the methods described in this chapter was made possible through grants from the National Institutes of Health, HD 04252, HD 00036, RR-05475, The National Foundation-March of Dimes, and the Chicago Community Trust. Serge B. Melancon is supported by the Quebec Medical Research Council and Henry L. Nadler is Given Research Professor of Pediatrics.

REFERENCES

1. H. L. Nadler, Prenatal detection of genetic defects, J. Pediat., 74, 132 (1969).

2. H. L. Nadler and A. B. Gerbie, The role of amniocentesis in the intra-uterine detection of genetic disorders, New Engl. J. Med., 282, 596 (1970).

3. H. L. Nadler, Prenatal detection of genetic disorders, in Advances in Human Genetics (H. Harris and K. H. Hirschhorn, eds.), Plenum, New York, 1972, pp. 1-37.

4. A. M. Milunsky, J. W. Littlefield, J. N. Kanfer, E. H. Kolodny, V. E. Shih, and L. Atkins, Prenatal genetic diagnosis, New Engl. J. Med., 283, 1370, 1441, 1498 (1970).

5. G. Wagner and F. Fuchs, The volume of amniotic fluid in the first half of human pregnancy, J. Obstet. Gynaecol. Brit. Commonwealth, 69, 131 (1962).

6. A. A. Plentl, Formation and circulation of amniotic fluid, Clin. Obstet. Gynecol., 9, 427 (1966).

7. H. E. Jacoby, Amniotic fluid volumes, Develop. Med. Child Neurol., 8, 587 (1966).

8. R. W. Bonsnes, Composition of amniotic fluid, Clin. Obstet. Gynecol., 9, 440 (1966).

9. D. R. Ostergard, The physiology and clinical importance of amniotic fluid, a review, Obstet. Gynecol. Survey, 25, 297 (1970).

10. A. R. Tankard, D. S. T. Bagnall, and F. Morris, The composition of amniotic fluid, Analyst, 59, 806 (1934).

11. R. E. Behrman, J. T. Parer, and C. W. de Lannoy, Jr., Placental growth and the formation of amniotic fluid, Nature, 214, 678 (1967).

12. A. C. Barnes, ed., Intrauterine Development, Lee and Febiger, Philadelphia, 1968.

13. T. M. Abbas and J. E. Tovey, Proteins of the liquor amnii, Brit. Med. J., 1960-1, 476.

14. J. F. D. Shrewsbury, Observations of the chemistry of liquor amnii, Lancet, 1933-I, 415.

15. A. Brzezinski, E. Sadovsky, and E. Shafrir, Protein composition of early amniotic fluid and fetal serum with a case of bis-albuminemia, Am. J. Obst. Gynecol., 89, 488 (1964).

16. M. Seppala, E. Ruoslahti, and T. H. Tallberg, Genetical evidence for maternal origin of amniotic fluid proteins, Ann. Med. Exptl. Biol. Fenniae (Helsinki), 44, 6 (1966).

17. J. Dancis, J. Lind, and P. Vera, in The Placental and Foetal Membranes (C. A. Villee, ed.), Williams & Wilkins, Baltimore, 1960, p. 185.

18. A. Brzezinski, E. Sadovsky, and E. Shafrir, Electrophoretic distribution of proteins in amniotic fluid and in maternal and fetal serum, Am. J. Obst. Gynecol., 82, 800 (1961).

19. J. F. Marks, J. Baum, J. L. Kay, W. Taylor, and L. Curry, Amniotic fluid concentrations of uric acid, Pediatrics, 42, 360 (1968).

20. C. W. Baird and I. E. Bush, Cortisone and cortisol contents of amniotic fluid from diabetic and non-diabetic women, Acta Endocrind., 34, 97 (1960).

21. T. N. A. Jeffcoate, J. R. H. Fliegner, S. H. Russell, J. C. Davis, and A. D. Wade, Diagnosis of the adrenogenital syndrome before birth, Lancet, 1965-II, 553.

22. I. R. Merkatz, M. I. New, R. E. Peterson, and M. P. Seaman, Prenatal diagnosis of adrenogenital syndrome by amniocentesis, J. Pediat., 75, 977 (1970).

23. A. I. Klopper and M. C. MacNaughton, The identification of preg-
    nanediol in liquor amnii, bile and faeces, J. Endocrinol., 18, 319
    (1959).

24. C. L. Cope, B. Hurlock, and C. Swell, The distribution of adrenal
    cortical hormone in some body fluids, Clin. Sci., 14, 25 (1955).

25. M. Lambert and G. W. Pennington, The estimation of polar steroids
    in liquor amnii, J. Endocrinol., 32, 287 (1965).

26. K. R. Abt and M. Keller, 17 Keto-steroids and phenolic steroids in
    amniotic fluid. Geburtsh. Frauenheilk., 14, 126 (1954).

27. F. A. Aleem, J. H. M. Pinerton, and D. W. Neill, Clinical signifi-
    cance of the amniotic fluid oestriol level, J. Obstet., Gynaecol. Brit.
    Commonwealth, 76, 200 (1969).

28. E. Diczfalusy and A. M. Magnusson, Tissue concentration of oestrone,
    oestradiol and oestriol in the human fetus, Acta Endocrinol., 28, 169
    (1958).

29. A. E. Schindler and W. L. Herrmann, Estriol in pregnancy urine and
    amniotic fluid, Am. J. Obstet. Gynecol., 95, 301 (1966).

30. A. E. Schindler, V. Ratanasopa, T. Y. Lee, and W. L. Herrmann,
    Estriol and Rh isoimmunization: A new approach to the management
    of severely affected pregnancies, Obstet. Gynecol., 29, 265 (1967).

31. P. Troen, B. Nilsson, N. Wizvist, and E. Diczfalusy, The pattern
    of estriol conjugates in normal human cord blood, amniotic fluid,
    and urine of newborns, Acta Endocrinol., 38, 371 (1961).

32. J. A. Bruner, Distribution of chorionic gonadotrophin in mother
    and fetus at various states of pregnancy, J. Clin. Endocrinol Metab.,
    11, 360 (1951).

33. T. Tallberg, E. Rouslakti, and C. Ehnholm, Immunological studies
    on human placental proteins and the purification of the human placental
    lactogen, Ann. Med. Exptl. Biol. Fenniae (Helsinki), 43, 67 (1965).

34. F. Fuchs, Discussion of paper by Jacobson and Barter, Am. J. Obst.
    Gynecol., 99, 806 (1967).

35. J. Nichols, Antenatal diagnosis and treatment of the adrenogenital
    syndrome, Lancet, 1970-I, 83.

36. F. Fuchs, Genetic information from amniotic fluid constituents,
    Clin. Obstet. Gynecol., 9, 565 (1966).

37. H. Alvarez, Diagnosis of hydatidiform mole by transabdominal placen-
    tal biopsy, Am. J. Obstet. Gynecol., 95, 538 (1966).

38. H. Sato and T. Kadotani, Fetal skin biopsy, JAMA, 212, 323 (1970).

39. A. W. Liley, The technique and complications of amniocentesis, New Zealand Med. J., 59, 581 (1960).

40. V. J. Freda, The Rh problem in obstetrics and a new concept of its management using amniocentesis and spectrophotometric scanning of amniotic fluid, Am. J. Obstet. Gynecol., 92, 341 (1965).

41. J. T. Queenan and D. W. Adams, Amniocentesis for prenatal diagnosis of erythroblastosis fetalis, Obstet. Gynecol., 25, 302 (1965).

42. J. T. Queenan, Amniocentesis and transamniotic fetal transfusion for Rh disease, Clin. Obstet. Gynec., 9, 491 (1966).

43. V. J. Freda, Recent obstetrical advances in the Rh problem, antepartum management, amniocentesis, and experience with hysterotomy and surgery in utero, Bull. N.Y. Acad. Med., 42, 474 (1966).

44. R. G. Burnett and W. R. Anderson, The hazards of amniocentesis, J. Iowa Med. Soc., 58, 130 (1968).

45. P. Riis and F. Fuchs, Antenatal determination of foetal sex in prevention of hereditary diseases, Lancet, 1960-II, 180.

46. C. R. Jacobson and R. H. Barter, Intrauterine diagnosis and management of genetic defects, Am. J. Obstet. Gynecol., 99, 796 (1967).

47. L. Van Leeuwen, H. Jacoby and D. Charles, Exfoliative cytology of amniotic fluid, Acta Cytol., 9, 442 (1965).

48. H. J. Huisjes, Origin of the cells in the liquor amnii, Am. J. Obstet. Gynecol., 106, 1222 (1970).

49. R. A. Votta, C. B. de Gagneten, O. Parada, and M. Giulietti, Cytologic study of amniotic fluid in pregnancy, Am. J. Obstet. Gynecol., 102, 571 (1968).

50. A. D. Hoyes, Ultrastructure of the cells of the amniotic fluid, J. Obstet. Gynaecol. Brit. Commonwealth, 75, 164 (1968).

51. E. Wachtel, H. Gordon, and E. Olsen, Cytology of amniotic fluid, J. Obstet. Gynaecol. Brit. Commonwealth, 76, 596 (1969).

52. F. Fuchs and P. Riis, Antenatal sex determination, Nature, 177, 330 (1956).

53. L. B. Shettles, Nuclear morphology of cells in amniotic fluid in relation to sex of infant, Am. J. Obst. Gynecol., 71, 834 (1956).

54. E. L. Makowski, K. A. Prem, and I. H. Kaiser, Detection of sex of fetuses by the incidence of sex chromatin body in nuclei of cells in amniotic fluid, Science, 123, 542 (1956).

55. D. M. Serr, L. Sachs, and M. Danon, Diagnosis of sex before birth using cells from amniotic fluid, Bull. Res. Council Israel, 5B137 (1955).

56. C. J. Dewhurst, Diagnosis of sex before birth, Lancet, 1956-I, 471.

57. F. James, Sexing foetuses by examination of the amniotic fluid, Lancet, 1956-I, 202.

58. E. Keymer, E. Silva-Inzunza, and W. E. Coutts, Contribution to the antenatal determination of sex, Am. J. Obstet. Gynecol., 74, 1098 (1957).

59. C. Pasquinucci, Studio della "chromatina sessuale" nelle cellule del liquido amniotico per la diagnosi prenatale di sesso, Ann. Ostet. Ginecol., 79, 152 (1957).

60. A. P. Amarose, A. J. Wallingford, and E. J. Plotz, Prediction of fetal sex from cytologic examination of amniotic fluid, New Eng. J. Med., 275, 715 (1966).

61. M. M. Nelson and A. E. H. Emery, Amniotic fluid cells: prenatal sex prediction and culture, Brit. Med. J., 1, 523 (1970).

62. D. M. Serr and E. Margolis, Diagnosis of fetal sex in a sex-linked hereditary disorder, Am. J. Obstet. Gynecol., 88, 230 (1964).

63. P. Riis and F. Fuchs, Sex chromatin and antenatal sex diagnosis, in The Sex Chromatin, (K. G. Moore, ed.), W. B. Saunders, Philadelphia, 1966.

64. R. P. Cox, G. Douglas, J. Hutzler, J. Lynfield and J. Dancis, In utero detection of Pompe's disease, Lancet, 1970-I, 893.

65. C. L. Y. Lee, N. M. Gregson, and S. Walker, Eliminating red blood cell from amniotic fluid samples, Lancet, 1970-II, 316.

66. M. C. Rattazzi and R. G. Davidson, Prenatal detection of Tay-Sachs disease, in Antenatal Diagnosis (A. Dorfman, ed.), University of Chicago, Chicago, 1972, p. 207.

67. S. B. Melancon, S. Y. Lee, and H. L. Nadler, Histidase activity in cultivated human amniotic fluid cells, Science, 173, 627 (1971).

68. C. De Duve, B. C. Pressman, R. Gianetto, R. Wattiaux, and F. Appleman, Tissue fractionation studies. 6 Intracellular distribution of enzymes in rat tissue, Biochem. J., 60, 604 (1955).

69. M. W. Steele and W. R. Breg, Chromosome analysis of human amniotic fluid cells, Lancet, 1966-I, 383.

70. A. Rook, L. Y. Hsu, M. Gertner, and K. Hirschhorn, Identification of Y and X chromosomes in amniotic fluid cells, Nature, 230, 53 (1971).

71. B. W. Uhlendorf, personal communication.

72. M. N. Macintyre, personal communication.

73. B. W. Uhlendorf and S. H. Mudd, Cystathionine synthase in tissue culture derived from human skin: enzyme defect in homocystinuria, Science, 160, 1007 (1968).

74. A. Dorfman, ed., Antenatal Diagnosis, University of Chicago, Chicago, 1972.

75. J. Dancis, J. Hutzler, and R. P. Cox, Enzyme defect in skin fibroblasts in intermittent branched-chain ketonuria and in maple syrup urine disease, Biochem. Med., 2, 407 (1969).

76. C. Orlandi, R. V. Torsello, and F. Bottiglioni, Analisi qualitativa e dosaggio semiquantitativo degli aminoacidi contenuti nel liquido amniotico, Attualita Ostet. Ginec., 4, 871 (1958).

77. Z. T. Wirtschafter, Free amino acids in human amniotic fluid, fetal and maternal serum, Am. J. Obstet. Gynecol., 76, 1219 (1958).

78. D. Sassi, Sulla presenza degli aminoacidi nel liquido amniotico, Monograph. Ostet. Ginec., 33, 683 (1962).

79. D. H. Spackman, Technicon Monograph No. 3, Geneva, p. 40 (1968).

80. H. L. Levy and P. P. Montag, Free amino acids in human amniotic fluid. A quantitative study by ion-exchange chromatography, Pediat. Res., 3, 113 (1969).

81. A. E. H. Emery, D. Burt, J. B. Scrimgeour, and M. M. Nelson, Antenatal diagnosis and the amino acid composition of amniotic fluid, Lancet, 1970-I, 307.

82. F. Cockburn, S. P. Robins, and J. O. Forfar, Free amino-acid concentrations in fetal fluids, Brit. Med. J., 1970-3, 747.

83. A. Saifer, E. A'Zary, and L. Schneck, Quantitative cation-exchange chromatographic analysis of free amino acids in human amniotic fluid collected during early pregnancy, Clin. Chem., 16, 891 (1970).

84. L. Dallaire and M. G. Gagnon, Etude semi-quantitative des acides amines du liquide amniotique, en relation avec l'age de la grossesse et les valeurs d'acides amines presents dans l'urine et le plasma maternels, Union Med. Canada, 100, 1116 (1971).

85. G. W. Frimpter, A. J. Greenberg, M. Hilgartner, and F. Fuchs, Cystathioninuria: management, Am. J. Diseases Children, 113, 115 (1967).

86. G. Morrow III, R. H. Schwarz, J. A. Hallock, and L. A. Barnes, Prenatal detection of methylmalonic acidemia, J. Pediat., 77, 120 (1970).

87. L. Kesner and E. Muntwyler, Automatic determination of weak organic acids by partition column chromatography and indicator titration, Anal. Chem., 38, 1164 (1968).

88. O. Stokke, L. Eldjarn, K. R. Norum, F. Steen-Johnsen, and S. Halvorsen, Methyl-malonic acidemia: a new inborn error of metabolism which may cause fatal acidosis in the neonatal period, Scand. J. Clin. Lab. Invest., 20, 31 (1967).

89. V. E. Shih and J. W. Littlefield, Argininosuccinase activity in amniotic fluid cells, Lancet, 1970-II, 45.

90. H. L. Nadler and A. B. Gerbie, Enzymes in noncultured amniotic fluid cells, Am. J. Obstet. Gynecol., 103, 710 (1969).

91. J. Dancis, The antepartum diagnosis of genetic diseases, J. Pediat., 72, 301 (1968).

92. Y. R. Hsia, K. J. Scully and L. E. Rosenberg, Inherited proprionyl-CoA carboxylase deficiency in "ketotic hyperglycinemia." Presented at the annual meeting of the American Pediatric Society and the Society for Pediatric Research, Atlantic City, April 29–May 2, 1970, p. 26.

93. V. E. Shih and J. D. Schulman, Ornithine-ketoacid transaminase activity in human skin and amniotic fluid cell culture, Clin. Chim. Acta, 27, 73 (1970).

94. J. A. Schneider, J. A. Rosenbloom, K. H. Bradley, and J. E. Seegmiller, Increased free cystine content of fibroblasts cultured from patients with cystinosis, Biochem. Biophys. Res. Commun., 29, 527 (1967).

95. K. Hummeler, B. A. Zajac, M. Genel, P. G. Holtzapple, and S. Segal, Human cystinosis: Intracellular deposition of cystine, Science, 168, 859 (1970).

96. R. O. Brady, J. N. Kanfer, and D. Shapiro, Metabolism of gluco-cerebrosides, II. Evidence of an enzymatic deficiency in Gaucher's disease, Biochem. Biophys. Res. Commun., 18, 221 (1965).

97. R. O. Brady, J. N. Kanfer, R. M. Bradley, and D. Shapiro, Demonstration of a deficiency of glucocerebroside cleaving enzyme in Gaucher's disease. J. Clin. Invest., 45, 1112 (1966).

98. E. Beutler, W. Kuhl, F. Trinidad, R. Teplitz, and H. Nadler, β-Glucosidase activity in fibroblasts from homozygotes and heterozygotes for Gaucher's disease, Am. J. Human Genet., 23, 62 (1971).

99. E. Beutler, W. Kuhl, F. Trinidad, R. Teplitz, and H. L. Nadler, Detection of Gaucher's disease and its carrier state from fibroblast cultures, Lancet 1970-II, 369.

100. O. H. Lowry, N. J. Rosebrough, A. L. Farr, and R. J. Randall, Protein measurment with the folin phenol reagent, J. Biol. Chem., 193, 265 (1951).

101. A. C. Crocker, The cerebral defect in Tay-Sachs disease and Niemann-Pick disease, J. Neurochem., 7, 69 (1961).

102. R. O. Brady, Genetics and the sphingolipidoses, Med. Clin. N. Am., 53, 827 (1969).

103. R. O. Brady, J. N. Kanfer, M. B. Mock, and D. S. Fredrickson, The metabolism of sphingomyelin: II. Evidence of an enzymatic deficiency in Niemann-Pick disease, Proc. Natl. Acad. Sci. U.S.A., 55, 366 (1966).

104. D. Shapiro and H. M. Flowers, Studies on sphingolipids, VII. Synthesis and configuration of natural sphingomyelins, J. Am. Chem. Soc., 84, 1047 (1962).

105. P. B. Schneider and E. P. Kennedy, Sphingomyelinase in normal spleens and from subjects with Niemann-Pick disease, J. Lipid Res., 8, 202 (1967).

106. H. R. Sloan, B. W. Uhlendorf, J. N. Kanfer, R. O. Brady, and D. S. Fredrickson, Deficiency of sphingomyelin-cleaving enzyme activity in tissue cultures derived from patients with Niemann-Pick disease, Biochem. Biophys. Res. Commun., 34, 582 (1969).

107. C. J. Epstein, R. O. Brady, E. L. Schneider, R. M. Bradley, and D. Shapiro, In utero diagnosis of Niemann-Pick disease, Am. J. Human Genet., 23, 533 (1971).

108. A. Holtz, B. W. Uhlendorf, and D. S. Fredrickson, Persistence of a lipid defect in tissue cultures derived from patients with Niemann-Pick disease, Federation Proc., 23, 128 (1964).

109. J. N. Kanfer, O. M. Young, D. Shapiro, and R. O. Brady, Metabolism of sphingomyelin, I. Purification and properties of a sphingomyelin-cleaving enzyme from rat liver, J. Biol. Chem., 241, 1081 (1966).

110. C. C. Sweeley, Purification and partial characterization of sphingomyelin from human plasma, J. Lipid. Res., 4, 402 (1963).

111. J. Austin, A. S. Balasubramanian, T. N. Pattabiraman, S. Saraswathi, D. K. Basu, and B. K. Bachhawat, A controlled study of enzymic activities in three human disorders of glycolipid metabolism, J. Neurochem., 10, 805 (1963).

112. E. Mehal and H. Jatzkewitz, Evidence for a genetic block in metachromatic leukodystrophy, Biochem. Biophys. Res. Commun., 19, 407 (1965).

113. J. Austin, D. Armstrong, and L. Shearer, Metachromatic form of diffuse cerebral sclerosis: V. The nature and significance of low sulfatase activity: a controlled study of brain, liver and kidney in four patients with metachromatic leukodystrophy (MLD), Arch. Neurol. (Chicago), 13, 593 (1965).

114. H. Jatzkewitz and E. Mehal, Cerebroside-sulfatase and aryl sulfatase A deficiency in metachromatic leukodystrophy (ML), J. Neurochem., 16, 19 (1969).

115. J. Austin, D. Armstrong, L. Shearer, and D. McAfee, Metachromatic form of diffuse cerebral sclerosis. VI A rapid test for sulfatase. A deficiency in metachromatic leukodystrophy urine, Arch. Neurol. (Chicago), 14, 259 (1966).

116. A. K. Percy and R. O. Brady, Metachromatic leukodystrophy: Diagnosis with samples of venous blood, Science, 161, 594 (1968).

117. J. V. Murphy, H. L. Wolfe, and H. L. W. Mosser, Multiple Sulfatase Deficiencies in a Variant Form of Metachromatic Leukodystrophy Lipid Storage Disease: Enzymatic defects and clinical implications (J. Bernsohn, ed.), Academic, New York (in press).

118. M. T. Porter, A. L. Fluharty, and H. Kihara, Metachromatic leukodystrophy: aryl sulfatase. A deficiency in skin fibroblast cultures, Proc. Natl. Acad. Sci., U.S.A., 62, 887 (1969).

119. M. M. Kaback and R. R. Howell, Infantile metachromatic leukodystrophy: Heterozygote detection in skin fibroblasts and possible applications to intrauterine diagnosis, New Engl. J. Med., 282, 1336 (1970).

120. R. O. Brady, A. E. Gal, R. M. Bradley, E. Martenson, A. L. Warshaw, and L. Laster, Enzymatic defect in Fabry's disease: ceramide trihexosidase deficiency, New Engl. J. Med., 276, 1163 (1967).

121. R. Matalon, A. Dorfman, G. Dawson, and C. C. Sweeley, Glycolipid and mucopolysaccharide abnormality in fibroblasts of Fabry's disease, Science, 164, 1522 (1969).

122. J. A. Kint, Fabry's disease: alpha-galactosidase deficiency, Science, 167, 1268 (1970).

123. C. A. Mapes, R. L. Anderson, and C. C. Sweeley, Galactosyl-galactosylglucoside ceramide: galactosyl hydrolase in normal human plasma and its absence in patients with Fabry's disease, FEBS Letters, 7, 180 (1970).

124. R. O. Brady, B. W. Uhlendorf, and C. B. Jacobson, Fabry's disease: antenatal detection, Science, 172, 174 (1971).

125. J. Austin, Studies in globoid (Krabbe) leukodystrophy. I. The significance of lipid abnormalities in white matter in 8 globoid and 13 control patients, Arch. Neurol. (Chicago), 9, 207 (1963).

126. B. K. Bachhawat, J. Austin, and D. Armstrong, A cerebroside sulfotransferase deficiency in a human disorder of myelin, Biochem. J., 104, 15c (1967).

127. D. Bowen and N. Radim, Cerebrosidegalactosidase: A method for determination and a comparison with other lysosomal enzymes in developing rat brain, J. Neurochem., 16, 501 (1969).

128. Y. Eto and K. Suzuki, Brain sphingoglycolipids in globoid-cell leukodystrophy. Meeting of the Am. Soc. for Neurochem., Albuquerque, New Mexico, March 16-18, 1970, p. 42.

129. M. Malone, Deficiency in a degradative enzyme system in globoid leukodystrophy. Meeting Am. Soc. Neurochem., Albuquerque, New Mexico, March 16-18, 1970, p. 42.

130. K. Suzuki, personal communication.

131. R. F. Shaw and A. P. Smith, Is Tay-Sachs disease increasing?, Nature, 224, 1213 (1969).

132. N. D. Myrianthopoulos and S. M. Aronson, Population dynamics of Tay-Sachs disease. Reproductive fitness and selection, Am. J. Human Genet., 18, 313 (1966).

133. E. H. Kolodny, R. O. Brady, and B. W. Volk, Demonstration of an alteration of ganglioside metabolism in Tay-Sachs disease, Biochem. Biophys. Res. Commun., 37, 526 (1969).

134. S. Okada and J. S. O'Brien, Tay-Sachs disease: generalized absence of a β-D-N-acetylhexosaminidase component, Science, 165, 698 (1969).

135. K. Sandhoff, Variations of β-N-acetyl hexosaminidase-pattern in Tay-Sachs disease, FEBS Letters, 4, 351 (1969).

136. C. G. Kolodny, B. S. Uhlendorf, J. M. Quirk, et al. Gangliosides in cultured skin fibroblasts accumulation in Tay-Sachs disease. 12th Intern. Congr. Biochem. Lipids, Athens, Greece, September 7-11, 1969, p. 39.

137. L. Schneck, J. Friedland, C. Valenti, M. Adachi, D. Amsterdam, and B. W. Volk, Prenatal diagnosis of Tay-Sachs disease, Lancet, 1970-I, 582.

138. K. Sandhoff, U. Andreae, and H. Jatzkewitz, Deficient hexosaminidase activity in an exceptional case of Tay-Sachs disease with additional storage of kidney globoside in visceral organs, Life Science, 7, 283 (1968).

139. Y. Suzuki, J. C. Jacob, and K. Suzuki, A case of G_{M2} gangliosidosis with total hexosaminidase deficiency, Neurology, 20, 388 (1970).

140. M. C. Rattazzi and R. G. Davidson, personal communication (manuscript in press).

141. J. Friedland, G. Perle, A. Saifer, L. Schneck, and B. W. Volk, Screening for Tay-Sachs disease in utero using amniotic fluid, Proc. Soc. Exptl. Biol. Med., 136, 1297 (1971).

142. J. O'Brien, S. Okada, A. Chen, and D. Fillerup, Tay-Sachs disease: Detection of heterozygotes and homozygotes by serum hexosaminidase assay, New Engl. J. Med., 238, 15 (1970).

143. J. O'Brien, Generalized gangliosidosis, J. Pediat., 75, 167 (1969).

144. K. Suzuki, K. Suzuki, and S. Kamoshita, Chemical pathology of G_{M1}-gangliosidosis (generalized gangliosidosis), J. Neuropathol. Exptl. Neurol., 28, 25 (1969).

145. L. S. Wolfe, J. Callahan, J. S. Fawcett, F. Andermann, and C. R. Scriver, G_{M1}-gangliosidosis without chondrodystrophy or visceromegaly, Neurology, 20, 23 (1970).

146. R. Sacrez, J. G. Juif, J. M. Gigonnet, and J. E. Gruner, La maladie de Landing ou idiotie amaurotique infantile precoce avec gangiosidose generalisee de type G_{M1}, Pédiatrie, 22, 143 (1967).

147. S. Okada and J. S. O'Brien, Generalized gangliosidosis: beta-galactosidase deficiency, Science, 160, 1002 (1968).

148. J. A. Kint, G. Dacremont, and R. Vlietinck, Type II G_{M1} gangliosidosis, Lancet, 1969-II, 108.

149. H. R. Sloan, B. W. Uhlendorf, C. B. Jacobson, and D. S. Fredrickson, β-galactosidase in tissue culture derived from human skin and bone marrow: Enzyme defect in G_{M1} gangliosidosis, Pediat. Res., 3, 532 (1969).

150. M. W. Ho and J. S. O'Brien, Hurler's syndrome: Deficiency of a specific beta galactosidase isoenzyme, Science, 165, 611 (1969).

151. P. A. Ockerman, Acid hydrolases in skin and plasma in gargoylism. Deficiency of beta-galactosidase in skin, Clin. Chim. Acta, 20, 1 (1968).

152. D. J. Yarborough, O. T. Meyer, A. M. Dannenberg, and B. Pearson, Histochemistry of macrophage hydrolases. III Studies on β-galactosidase, β-glucuronidase and aminopeptidase with indolyl and naphtyl substrates, J. Reticuloendothel. Soc., 4, 390 (1967).

153. E. Klenk and W. Kahlke, Uber das vorkommen der 3.7.11.15-tetramethyl hexadecansaure (phytansaure) in den cholesterinestern und anderen lipoidfraktionen der organe bei einem krankheitsfau unbekannter genese (verdacht auf heredopathia atactica polyneuritiformis-Refsum's syndrome), Hoppe-Seylers Z. Physiol. Chem., 333, 133 (1963).

154. J. H. Herndon, D. Steinberg, B. W. Uhlendorf, and H. M. Fales, Refsum's disease: Characterization of the enzyme defect in cell cultures, J. Clin. Invest., 48, 1017 (1969).

155. C. E. Mize, J. Avigan, J. H. Baxter, H. M. Fales, and D. Steinberg, Metabolism of phytol-U-^{14}C and phytanic acid-U-^{14}C in the rat, J. Lipid Res., 7, 692 (1966).

156. A. Karmen, A. Guiffrida, and R. W. Bowman, Radioassay by gas liquid chromatography of lipids labeled with carbon-14, J. Lipid Res., 3, 44 (1962).

157. J. C. Fratantoni, E. F. Neufeld, B. W. Uhlendorf, and C. B. Jacobson, Intrauterine diagnosis of the Hurler and Hunter syndromes, New Engl. J. Med., 280, 686 (1969).

158. R. Matalon, A. Dorfman, H. L. Nadler, and C. B. Jacobson, A chemical method for the antenatal diagnosis of mucopolysaccharidoses, Lancet 1970-I, 83.

159. B. S. Danes and A. G. Bearn, Hurler's syndrome, a genetic study in cell culture, J. Exptl. Med., 123, 1 (1966).

160. B. Sylven, On the interaction between metachromatic dyes and various substrates of biological interest, Acta Histochem. Suppl., 1, 79 (1958).

161. A. Milunsky and J. W. Littlefield, Diagnostic limitations of metachromasia, New Engl. J. Med., 281, 1128 (1969).

162. A. Dorfman and R. Matalon, The Hurler and Hunter syndromes, Am. J. Med., 47, 691 (1969).

163. B. S. Danes, J. E. Scott, and A. G. Bearn, Further studies on metachromasia in cultured human fibroblasts. Staining of glycosaminoglycans (mucopolysaccharides) by Alcian blue in salt solutions, J. Exptl. Med., 132, 765 (1970).

164. J. C. Fratantoni, C. W. Hall, and E. F. Neufeld, The defect in Hurler's and Hunter's syndromes: Faulty degradation of mucopolysaccharides, Proc. Natl. Acad. Sci. U.S.A., 60, 699 (1968).

165. J. D. Gregory and P. W. Robbins, Metabolism of sulfur compounds (sulfate metabolism), Ann. Rev. Biochem., 29, 347 (1960).

166. V. Magrini, M. Fraccaro, L. Tiepolo, S. Scappaticci, L. Lenzi, and G. P. Perona, Mucopolysaccharidoses: Autoradiographic study of sulfate $-^{35}$S uptake by cultured fibroblasts, Ann. Humun. Genet. London, 31, 231 (1967).

167. G. M. Komrower, Galactosemia, Proc. Roy. Soc. Med., 60, 1155 (1967).

168. H. L. Nadler, C. M. Chacko, and M. Rachmeler, Interallelic complementation in hybrid cells derived from human diploid strains deficient in galactose-1-phosphate uridyl transferase activity, Proc. Natl. Acad. Sci. U.S.A., 67, 976 (1970).

169. G. Hug and W. K. Schubert, Glycogenosis type II. glycogen distribution in tissues, Arch. Pathol., 84, 141 (1967).

170. H. G. Hers, Glucosidase deficiency in generalized glycogen-storage disease (Pompe's disease), Biochem. J., 86, 11 (1963).

171. H. L. Nadler and A. M. Messina, In-utero detection of Type II Glycogenosis (Pompe's disease), Lancet, 1969-II, 1277.

172. R. P. Cox, G. Douglas, J. Hutzler, J. Lynfield, and J. Dancis, In-utero detection of Pompe's disease, Lancet, 1970-I, 893 (1970).

173. H. L. Nadler, R. H. Bigley, and G. Hug, Prenatal detection of Pompe's disease, Lancet, 1970-II, 369.

174. I. S. Salafsky and H. L. Nadler, α-1,4 Glucosidase activity in Pompe's disease, J. Pediat., 79, 794 (1971).

175. H. M. Nitowsky and A. Grunfeld, Lysosomal glucosidase in Type II glycogenosis activity in leukocytes and cell cultures in relation to genotype, J. Lab. Clin. Med., 69, 472 (1967).

176. A. Dahlquist, Determination of maltase and isomaltase activities with a glucose-oxidase reagent, Biochem. J., 80, 547 (1961).

177. P. Durand, C. Borrone, and G. Della Cella, Fucosidosis, J. Pediat., 75, 665 (1969).

178. F. Van Hoof and H. G. Hers, Mucopolysaccharidosis by absence of α-fucosidase, Lancet, 1968-I, 1198.

179. F. Van Hoof and H. G. Hers, The abnormalities of lysosomal enzymes in mucopolysaccharidoses, European J. Biochem. 7, 34 (1968).

180. P. A. Ockerman, Mannosidosis: Isolation of oligosaccharide storage material from brain, J. Pediat., 75, 360 (1969).

181. B. Kjellman, I. Gamstorp, A. Brun, P. A. Öckerman, and B. Palmgren, Mannosidosis: a clinical and histopathologic study, J. Pediat., 75, 366 (1969).

182. M. Lesch and W. L. Nyhan, A familial disorder of uric acid metabolism and central nervous system function, Am. J. Med., 36, 561 (1964).

183. J. C. Seegmiller, F. M. Rosenbloom, and W. N. Kelley, Enzyme defect associated with sex-linked human neurological disorder and excessive purine synthesis, Science, 155, 1682 (1967).

184. W. N. Kelley, M. L. Greene, F. M. Rosenbloom, J. F. Henderson, and J. E. Seegmiller, Hypoxanthine-guanine phosphoribosyltransferase deficiency in gout, Ann. Internal Med., 70, 155 (1969).

185. J. A. McDonald and W. N. Kelley, Lesch-Nyhan syndrome: Altered kinetic properties of mutant enzyme, Science, 171, 689 (1971).

186. P. H. Berman, M. E. Balis, and J. Dancis, A method for the prenatal diagnosis of congenital hyperuricemia, J. Pediat., 75, 488 (1969).

187. J. Dancis, R. P. Cox, P. H. Berman, V. Jansen, and M. E. Balis, Cell population density and phenotypic expression of tissue culture fibroblasts from heterozygotes of Lesch-Nyhan's disease (inosinate pyrophosphorylase deficiency), Biochem. Genet., 3, 609 (1969).

188. W. Y. Fujimoto and J. E. Seegmiller, Hypoxanthine-guanine phosphoribosyltransferase deficiency: Activity in normal, mutant, and heterozygote cultured human skin fibroblasts, Proc. Natl. Acad. Sci. U.S.A., 65, 577 (1970).

189. W. Y. Fujimoto, J. E. Seegmiller, B. W. Uhlendorf, and C. B. Jacobson, Biochemical diagnosis of an X-linked disease in utero, Lancet, 1968-II, 511 (1968).

190. R. DeMars, G. Sarto, J. S. Felix, and P. Benke, Lesch-Nyhan mutation: Prenatal detection with amniotic fluid cells, Science, 164, 1303 (1969).

191. J. F. Marks, J. Baum, J. L. Kay, W. Taylor, and L. Curry, Amniotic fluid concentrations of uric acid, Pediatrics, 42, 360 (1968).

192. G. A. Bray, A simple efficient liquid scintillator for counting aqueous solutions in a liquid scintillation counter, Anal. Biochem., 1, 279 (1960).

193. H. L. Nadler and T. J. Egan, Deficiency of lysosomal acid phosphatase: A new familial metabolic disorder, New Engl. J. Med., 282, 302 (1970).

194. O. H. Lowry, in Methods in Enzymology, Vol. 4 (S. P. Colowick and N. O. Kaylan, eds.), Academic, New York, 1959, p. 371.

195. C. H. Fiske and Y. Subbarow, Colorimetric determination of phosphorus, J. Biol. Chem., 66, 375 (1925).

196. A. G. E. Pearse, in Histochemistry, Vol. I, Little, Brown & Co., Boston, 1968, p. 729.

197. J. G. B. Russell, Radiology in the diagnosis of fetal abnormalities, J. Obstet. Gynaecol. Brit. Commonwealth., 76, 345 (1969).

198. J. T. Queenan and E. Gadow, Amniography for detection of congenital malformations, Obstet. Gynecol., 35, 648 (1970).

199. O. Aguero and I. Zighelboim, Fetography and molegraphy, Surg. Gynecol. Obstet., 130, 649 (1970).

200. J. Erbsloh, Das intra-uterine fetogramin, Arch. J. Gynak., 173, 160 (1942).

201. J. Jarousse, Resultats d'une serie d'amnioscopies, Bull. Fed. Soc. Gynecol. Obstet. Franc., 20, 370 (1968).

202. R. D. Ionascu, Examens comparatifs par amnioscopie et par amniocentese dans l'érythroblastose foetale, par isoimmunisation anti-Rhesus, Fed. des Soc. de Gynec. et d'Obstet. de langue Francaise (Masson et Cie, ed.), Paris, 1968, p. 421.

203. W. J. Garrett, G. Grunwald, and D. E. Robinson, Prenatal diagnosis of fetal polycystic kidney by ultrasound, Australia, New Zealand, J. Obstet. Gynaecol., 10, 7 (1970).

204. F. Pauls and P. Boutros, The value of placental localization prior to amniocentesis, Obstet. Gynecol., 35, 175 (1970).

205. K. R. Gottesfeld, H. E. Thompson, J. H. Holmes, and E. S. Taylor, Ultrasonic placentography - a new method for placental localization, Am. J. Obstet. Gynecol., 96, 538 (1966).

206. I. Donald, Sonar as a method of studying prenatal development, J. Pediat., 75, 326 (1969).

Chapter 2

BEHAVIORAL METHODS IN PHARMACOLOGY

Lewis S. Seiden[†]
University of Chicago
Departments of Pharmacology and Psychiatry
Chicago, Illinois

[†] Supported by Research Career Development Award 5-K2-MH-10, 562.

I. INTRODUCTION

The relatively young discipline of <u>behavioral pharmacology</u> studies the effects of drugs on behavior and their mechanism of action, using techniques from both experimental psychology and pharmacology. Experimentation commonly involves the definition of a behavioral task; all parameters controlling performance, with the exception of drug administration, are held constant during both the control and experimental periods. This is the so-called "N+1" design, where N is equal to the sum of all parameters in the control period. This philosophy of experimental design is common to many experimental sciences. The concept of <u>control</u> and measurement of <u>change from control</u> during the systematic variation of a single parameter (independent variable) is common to many scientific investigations.

If the effects of drugs on behavior are to be assessed, it is important to have a known prior baseline against which drug performance can be compared. The first problem of the experimenter, then, becomes one of establishing experimental control of the animal's behavior. This is done by imposing environmental conditions on the animal in order to generate stable and predictable behavior suitable for experimental work. A convenient means of establishing experimental control of behavior is through the use of operant conditioning in which the behavior of the organism is controlled primarily by the consequences produced by the organism's behavior. The examples of operant procedure outside the laboratory are many. Farmer Jones plants seeds and gets corn. Smith makes wheels and sells them for money, etc. The simple point is that the behavior produces consequences (corn, money) and these consequences are capable of predictably maintaining behavior over extended periods of time (planting corn, building wheels). This method of controlling behavior is a very powerful technique of generating reproducible behavior in the laboratory under specified experimental conditions; it has also been used in the analysis of the effects of drugs on behavior. The basic functional relationships between different variables that influence operant behavior are discussed with a view toward examining how these procedures can be applied to the analysis of drug action on behavior.

Systematic analysis of behavior has often been separated into two different paradigms: (1) <u>respondent behavior</u> (Pavlov) and (2) <u>operant behavior</u> (Skinner). In the following discussion the terms <u>stimulus</u> and <u>response</u> are used frequently. A response may be defined as a class of behavior; examples of simple responses are: flexing a leg within certain limits, salivation in a specified quantity, pressing a lever with force sufficient to operate it. Flexion and extension of various muscles involved in the production of such responses would not by themselves be considered a response but could be defined as such. Stimuli define other events, usually physical, and usually presented by the experimenter. For the purpose of this discussion the definition of a stimulus is limited to those physical events that are related to a response. While this chapter deals mainly with operant behavior, both operant and respondent behavior are briefly compared to clearly delineate their important features.

While studying the physiology of gastric secretions, Pavlov prepared dogs with gastro-intestinal fistulae. He then presented different types of food and measured the rate of secretion of gastric juices. The procedure followed a daily routine: Pavlov would enter the room and prepare the animal for the experiment; next, the animal caretaker would appear with the food. Pavlov noted that after several days the dog would begin to se-crete gastric juices when the caretaker entered. Pavlov later defined food as an <u>unconditioned stimulus</u> <u>(UCS)</u> in that it "naturally" elicited the gastric response. The caretaker on the other hand, had become a <u>conditioned</u> <u>stimulus</u> <u>(CS)</u>; after the animal's training (conditioning) the caretaker now elicited the same response (salivation) as the UCS. Behavior which can be elicited by a clearly identifiable stimulus or set of stimuli initially has the following model:

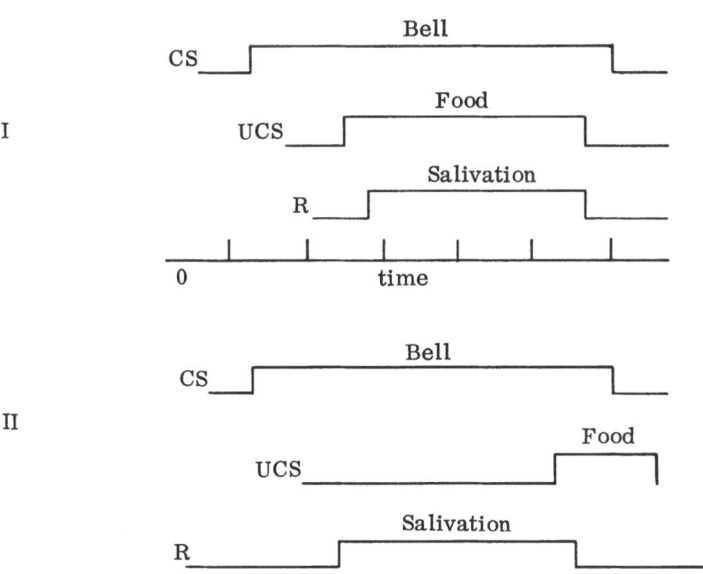

In this and the following diagrams an upward inflection of the line de-noted onset of the stimulus or response and a downward deflection denotes offset. CS is a conditioned stimulus, UCS is an unconditioned stimulus, and R is the response. By definition, the UCS elicits R; the CS <u>initially</u>

does not. When the CS (e.g., bell) and the UCS (e.g., food in the mouth)
are repeatedly presented in the temporal relationship described above, the
CS begins to elicit the response that initially was elicited by the UCS alone.
The constraints of stimulus duration, the temporal juxtaposition of the
stimuli, and their relative intensity influence the development and maintenance of a conditioned response (1).

Many psychologists attempted to use the respondent model to explain
all behavior, but B. F. Skinner (2) observed that it was extremely awkward
to explain all behavior in terms of eliciting stimuli. Skinner observed that
frequently it was more logical to describe the relationship between the
behavior and its consequences rather than the stimuli that elicited it. For
example, it makes more sense to say that Mr. Smith goes to work every
morning for the weekly paycheck than to analyze this complex behavior as
a series of respondents (i.e., responses elicited by stimuli) with the money,
or perhaps what the money will buy, as the final UCS. Skinner developed
principles that control operant behavior: the set of responses (pressing a
lever, turning a wheel, running a maze, etc.) whose rate of occurrence is
governed by their consequences (e.g., lever pressing leads to food or
water; running a maze to escape from shock).

In operant conditioning the response by the animal and the rate and
form of the response are modified by its resulting consequence. In respondent conditioning, the controlling events precede the response, which
is elicited by them. The operant response (or more simply, the operant)
is maintained by its consequences, that is, by the particular set of stimuli
(reinforcement) presented when a particular response occurs. An example
of the difference between operant and respondent conditioning is found in
the control of salivation where the bell and the fool are the controlling variables, and salivation is the respondent. But where the control over salivation is by consequences, it is operant, as when salivation indicates con-
tempt, or where fines and other punitive consequences are used to inhibit
it. In addition to the formal analytic differences between respondent and
operant conditioning there are procedural differences for establishing and
maintaining their control.

The following section defines terms and procedures involved in the
experimental analysis of operant behavior; it also includes data analysis
and application to drug research as well as factors which can affect the
interpretation of data. The area of operant behavior has proved useful in
many fields of endeavor and constitutes a highly sophisticated and complex
body of ideas, methods, and principles (2-8). It would be foolhardy to
attempt to review all fundamental experimental methods or the various
areas of application of operant techniques. Therefore, this section focuses
mainly on the methods and concepts essential for understanding operant
techniques found in the literature of behavioral pharmacology.

II. OPERANT BEHAVIOR

A. Terms

1. Reinforcement

Since operant behavior has been defined as a set of responses maintained by stimuli whose presentation is contingent on them, the nature of the stimuli becomes critical. If a stimulus presented immediately after a response is sufficient to affect the rate of the response, the stimulus is termed a reinforcer. There are two classes of reinforcers: appetitive and aversive. Appetitive reinforcement (e.g., food, water, money, etc.) and aversive reinforcement (painful stimuli, such as shock) will be further defined and discussed in the following section (See Table 1).

TABLE 1

Definitions of Types of Reinforcers in Terms of Response Rates

Response contingencies	Reinforcer may be defined as appetitive (S^+) if:	Reinforcer may be defined as aversive (S^-) if:
R ⟶ S (Response leads to the occurrence of a stimulus)	The contingencies can lead to an increase in the rate of responding: e.g., R ⟶ food R ⟶ water (positive reinforcer)	The contingencies can lead to a decrease in the rate of responding: e.g., R ⟶ shock (punishment)
	and	and
R ⟶ S (Response leads to the removal of a stimulus)	The contingencies lead to a decrease in the rate of responding: e.g., R ⟶ no food R ⟶ no water This particular set of contingencies is referred to as extinction (punishment)	The contingencies lead to an increase in response rate: R ⟶ termination of shock; as in escape R ⟶ postponement of shock or conditions associated with shock as in avoidance (negative reinforcer)

2. Appetitive

Consider a rat deprived of water for 23 hours and placed into an apparatus which releases a small quantity of water when the rat presses a lever. Under these circumstances, the depression of the lever is defined as an <u>operant response</u>, and the water would be considered <u>reinforcement</u>. In the scheme of the operant paradigm ($\underline{9},\underline{10}$) (see Fig. 1), this would be noted as:

$$R \longrightarrow S^+$$

where R is the response and S is the stimulus contingent on the response; S^+ may be food, water, intracranial stimulation, etc. An apparatus in which the closure of a microswitch leads to the delivery of reinforcement is the operant chamber designed by Skinner ($\underline{2}$) and is referred to as a "Skinner Box." In the operant chamber for pigeons (Fig. 2) reinforcement occurs whenever the subject closes a circuit through a lever-activated microswitch. The particular muscle movements or response topology that lead to closure of the switch are usually not specified.

<u>Appetitive reinforcement</u> is defined as a stimulus resulting from a behavioral response (e.g., lever press), the presentation of which will lead to an increase in the frequency of a response. Failure of the appetitive reinforcer to be presented when the $R \longrightarrow S^+$ chain is established will, in the long run, lead to a decrease in the rate of response (i.e., extinction). The conditions under which certain stimuli may acquire appetitive reinforcing properties are achieved by certain antecedent manipulations such as deprivation. These manipulations are called potentiating variables (PV) in the Goldiamond notation (Fig. 1). It is often said that "you can lead a horse to water but you can't make him drink." Strictly speaking this is not true. If the horse is deprived of water for a long enough period of time there is a high probability that he will drink. On the other hand, if he has

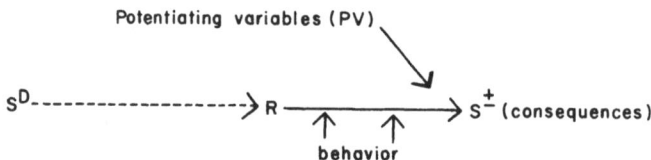

FIG. 1. The operant paradigm (Goldiamond, 1971). A diagrammatic representation of the factors influencing the contingency relationship between a response and reinforcement. S^D: a discriminative stimulus which signals the availability of reinforcement. R: the behavioral response. $S^\pm$: the reinforcing stimulus. PV: the potentiating variable.

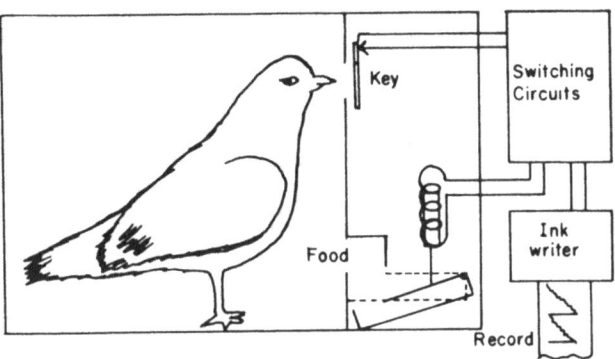

FIG. 2. Diagram of Skinner box. [Reproduced from an article by Peter B. Dews in the Journal of Pharmacology and Experimental Therapeutics, 115, 393–401 (1955) with permission from the Williams and Wilkins Company.]

been tied to the trough all day, one cannot predict his drinking behavior at any given moment. The point in this is that the antecedent conditions of the availability of a reinforcer will determine its effectiveness.

B. Shaping an Operant Response

The process of "shaping" a response (such as lever pressing) is facilitated by a program which begins by shaping the terminal drinking behavior and then proceeds to shape the desired operant by small steps. One type of program which is very effective in shaping a rat to press a lever for water reinforcement involves three steps to be described below.

1. Water Deprivation

Since in this case the desired behavior consists of a response, lever pressing, that will be reinforced with water, it is necessary to deprive the rat of water. This is best done by beginning deprivation three or four days prior to training. The water bottles are removed from the cages on day 1 of the deprivation phase of the program. Thereafter, the rat is allowed access to water once every 24 hours for about 5–15 minutes. At first, the time (latency) for returning to the water bottle (or food) is long, but within two or three days the animal begins to drink as soon as the water bottle is placed on the cage. It is important to weigh the animals each day during deprivation to ascertain that deprivation is not too severe. Animals should not fall below 70% of their body weight when allowed full access to food and

water. Recording body weights and food or water consumption gives a
check on the effectiveness of the deprivation conditions. In shaping procedures it is important that deprivation be sufficiently long and that the
size of the reinforcer be of the proper magnitude. In the case of the rat,
it has been found that it is very effective to initiate deprivation four or five
days before training and present the reinforcer (food or water) in the home
cage on each day at the time when the rat will be tested in order to establish
a rhythm of feeding or drinking. Too much reinforcement in operant training will cause satiation and a loss of stimulus (reinforcement) control.

2. Magazine Training

The deprived rat is placed into the box with the delivery cup filled with
water (0.01 ml to 1.0 ml). The rat usually discovers and consumes water
rapidly. In this phase of training, the experimenter can deliver water into
the cup by manually activating a switch; this procedure not only fills the
cup but produces a characteristic click. Through repeated presentations
of water and clicks the rat quickly approaches the cup in response to the
click. In this situation, the click is referred to as an S^D (discriminative
stimulus).

3. "Shaping" of Lever Pressing

When approach in response to the click is established, shaping of the
lever pressing response may begin by making the click and delivery of
water contingent upon movements similar to the lever pressing response.
The experimenter can readily do this by consistently reinforcing approaches
to the lever, by progressively raising the criterion for reinforcement: a
common sequence is the animal looking at the lever, touching the lever,
and pressing the lever.

While it is necessary to include water deprivation in any shaping program it is possible to combine magazine training and shaping of the lever
pressing response into one step. Moreover, under certain conditions experimenter intervention is not necessary and the experimental animal when
sufficiently deprived will learn through trial and error.

C. Schedule of Appetitive Reinforcement

Under conditions of positive reinforcement the rate of responding is
generally dependent on the amount of deprivation and density of reinforcement (PV and S^+ in Fig. 1). Reinforcements can be programmed to occur
intermittently instead of occurring after every lever press; the schedule of
reinforcement will have an effect on the rate of responding as well as the
distribution of responding over time. A schedule of reinforcement in which

the reinforcement occurs after every lever press is referred to as a continuous reinforcement schedule (CRF). Using a CRF schedule, the amount
of reinforcement can have a significant effect on the PV. Therefore, when
longer runs are desired, schedules of intermittent reinforcement are used.
In addition, each schedule of reinforcement generates a rate and distrubu-
tion of responding which is characteristic of that schedule; as we shall see,
both the rate of responding and pattern of responding are critically important determinants of the effects of drugs. The basic schedules that are
most often used in measuring drug effects have been reviewed. There are,
of course, countless combinations and parametric variations of these
schedules.   Table 2 (6) shows the relative response rates and distributions for the basic schedules of reinforcement.

1. Ratio Schedules

a. Fixed Ratio (FR). Under this schedule every Nth response in a
series will be reinforced. FR-10 indicates that the tenth response in a
chain of responses will be reinforced. On an FR schedule responses occur
in rapid bursts until reinforcement occurs, after which there is a post-reinforcement pause (Fig. 3). Within limits, the response rate on an FR
schedule will be proportional to the ratio; the response rate under FR 10
contingencies will be higher than under FR 2. However, higher rates
produced by longer ratios will remain high even when the ratio requirement is returned to a lower value. Therefore, if the experimenter is interested in examining rate-dependent action of drugs by using different
FRs, he must start with the lower ratios and proceed to the higher ones.

Experimenters have found that FR schedules generate characteristic
patterns of responding. Sometimes FR responding is characterized by
bursts of responses at the normal FR rate intermixed with long pauses
which are unrelated to reinforcement. This is referred to as FR straining. The size of a fixed ratio which can be conditioned on an FR schedule
depends on the species of animal, the nature of the operant (weight and
position of the lever), magnitude of PV, and the conditioning history of
the animal. Performance on an FR schedule may be developed from a CRF
schedule provided that the progression of the ratio is not too large; otherwise, in changing from a CRF schedule to a high FR, one may encounter
extinction. With high FR schedules it is best to have several intermediate
ratios to avoid extinction or straining.

b. Variable Ratio (VR). Under this schedule, reinforcement occurs
after N responses have been emitted, but N varies randomly over a certain
range of values. Thus, for a VR 10 schedule, the average value of the
ratio is 10, but reinforcement could occur after 2, 4, 5, 8, 10, 12, 14...
or 18 presses. Such particular values of the VR vary at random. As an

TABLE 2

Characteristics of Schedules of Reinforcement, Long Historical

Schedule	Contingencies	Conditioning	Extinction
Continuous (CRF)	Every R produces S^R	High, steady rate	High initial rate; low R; abort time
Fixed Ratio (FR)	Every nth R produces S^R (FR 50)	High rates; pauses increase with size of ratio	Fairly rapid; R omitted early at high rate. Any later R occur at high rate.
Fixed Interval (FI)	First R following a designated interval of time is reinforced (FI 10 min)	Low rate, longer scallops with longer intervals	Initial scallop; low sustained rate with occassional scallop appropriate to interval
Variable Interval (VI)	An interval schedule in which intervals between S^R vary at random about some mean (VI 5 min)	No pauses or scallops; fairly low rate	Sustained; response gradually tapers off
Variable Ratio (VR)	S^R occurs after a given number of R which varies about some mean (VR 50)	High, sustained rate	Fairly rapid. Most R emitted early at high rate. Many R.
Differential Reinf. Ratio drl (low)	drl; a pause of specified length must occur before S^R (drl 6 sec)	Low rates	Smooth curve, uniform low rate

drh (high)	drh; rate must reach some value before S^R (drh 2 per sec)	High rates	Periods of sustained high rates alternate with pauses
Pacing	Both upper and lower limits of rate specified		
Multiple (mult)	2 or more schedules with different S^D present during each. S^R follows each. Random order. (mult FR 100 FI 10)	Characteristics of component schedules	At first multiple control maintained, then diminishes
Mixed (mix)	2 or more schedules but no S^D correlated with schedule. S^R follows each. Random order. (mix FR 100 FI 10)	Maintained more easily than multiple	Characteristics of mixed schedule present
Chained (chain)	R or I schedule in presence of an S^D produces new S^D after which R or another schedule brings S^R (chain VI 3 FI 3)	FI VI easier then VI FI; FI FR easier than FR FI (to get characteristic curves)	
Tandem (tand)	A single S^R programmed by 2 schedules acting in succession without correlated S^Ds. (tand FI 45 FR 10)	Tend to get features of both components	Retains features of tandem. FI 45 FR 10 has higher rate, longer ext. than plain FI 45.
Concurrent (concurrent)	2 or more responses (keys). Can be reinforced on different schedules (concurrent FI 5 FR 50)	Since 2 R possible (but not at same time) separate curves reciprocal	

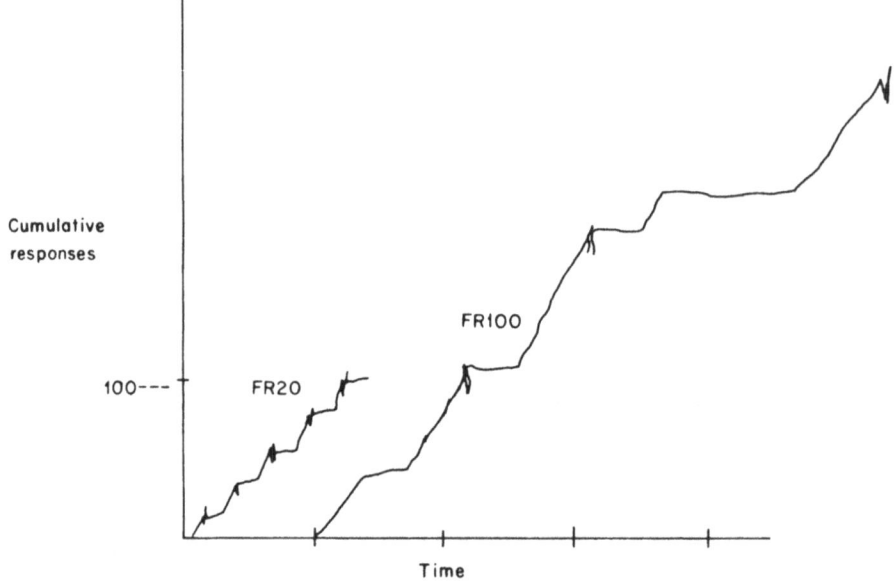

FIG. 3. Cumulative record illustrating FR straining. FR 20 record show sample of record in which the slashes indicated the delivery of an appetitive reinforcer. Not that the only "pauses" in the record are those occurring after reinforcement. In the FR 100 record, the post-reinforcement pauses are increases and pauses on this high ratio occur in the middle of the FR run. This is referred to as FR straining. Further increases in the ratio can lead to extinction.

example, the rat may press 4 times and be reinforced; on the next trial it may take 18 presses, and on the next 10, and so on. On a VR 10 the average number of presses would be 10 for reinforcement but a range of values around 10 would also lead to reinforcement.

2. Interval Schedules

a. Fixed Interval (FI). The first response that occurs after a fixed length of time has elapsed is reinforced. This schedule engenders a period of low rates of responding which is followed by higher rates as the time for reinforcement approaches (Table 2). On an FI schedule, responding begins well before the time that a single press delivers reinforcement. Since the responses made before the interval has elapsed do not result in reinforcement, this pattern of responding is often thought of as an example of superstitious behavior. Under most FI schedules the only discriminative stimulus is time. However, experimenters have presented stimuli (a sequential pattern of lights) which act as an S^D for the availability of reinforcement. Such stimuli can serve as a clock. Laties et al. (11) have studied the function of mediating stimuli and found that the opportunity to engage in mediating behaviors or the presence of time-correlated discriminative stimuli facilitate the development of sharply discriminate FI responding.

b. Variable Interval (VI). Responses on this schedule are intermittently reinforced over time. For example, on a VI 1 (variable interval 1 min) responding is reinforced once per minute (mean reinforcement interval) on the average; some reinforcement intervals may occur at 10 sec (minimum reinforcement interval) while others may occur at 120 sec or longer. The VI schedule engenders a uniform response rate. Since the density of reinforcements tends to be low, satiation does not occur and responding will persist for a longer period of time. Therefore, VI schedules are highly resistant to extinction. The response rate is dependent on both the mean reinforcement interval and the minimum reinforcement interval, and as the mean and minimum interval becomes shorter the response rate increases.

In the VI as in other schedules (e.g., FI) it is possible to influence the rate of responding by placing other constraints on the schedule. With a limited hold a potential reinforcement is held for only a certain time.

c. Differential Reinforcement of Low Rate (DRL). Up to this point the primary measurements of responding considered have been the rate and distribution of responding. The time interval between any two responses is referred to as an interresponse time (IRT). IRT measurement is often a very useful parameter for the measurement of responding and is defined as the time interval between the Nth response and the N + 1 response. It is the reciprocal of the instantaneous rate, i.e.,

(1) rate = responses/unit time

(2) IRT = time/responses

It follows from these definitions that the unit of time in (1) will be defined by time elapsed between any two responses. In this procedure, IRTs must be a certain interval apart in order for S^D conditions to be maintained. By manipulating this interval, either high or low rates of response can be achieved (12).

When a DRL schedule is in effect, reinforcement occurs if, and only if, the responses are separated by a certain minimal time. For example, a DRL of 10 sec indicates that only IRTs of 10 sec or longer are reinforced. IRTs shorter than 10 sec reset a clock which must count down before a lever press produces reinforcement and another 10-sec period without a response must elapse before reinforcement can occur. A DRL schedule engenders low rates of responding with an even distribution of IRTs clustered around a mean value close to the minimum time for reinforcement.

3. Schedules Employing More than One Schedule of Reinforcement

These schedules constitute the main types of positive reinforcement used in the study of the effects of drugs on operant behavior. Frequently, however, an investigator may desire to examine the effects of a drug on more than one schedule of reinforcement. It is possible to train an animal to emit behavior appropriate to more than one schedule. In a multiple schedule, two or more schedules are presented. Each is signaled by a different stimulus called an S^D. For example, one could set up a situation in which during the presence of a red light an FI 1 is in effect, and during the presence of a blue light an FR 10 is in effect. The rat with repeated experience with the S^Ds, learns to emit the appropriate rate and pattern.

A mixed schedule, similar to the multiple schedule has two or more programs associated with it but there are no discriminative stimuli. The FI alternates with the FR 10; after running off ten responses to reach FR 10 the rat would then go to an FI 1, etc. With a chain schedule two or more schedules are used and an S^D is correlated only with one, the one that leads to reinforcement. With a chained FI 1/FR 10, the equipment would be programmed on an FI 1 initially; the first press after a minute had elapsed would change the stimulus conditions (e.g., lights out) at which time the program would require ten presses for reinforcement. After reinforcement the S^D disappears. A tandem schedule is one in which two schedules sequentially lead to a single reinforcement, but there are no discriminative stimuli. For a more detailed description of response characteristics under each schedule as well as methods of shaping different schedules Refs. 4 and 6 may be consulted.

D. Aversive Reinforcement

The application of punishment will by definition cause a decrease in the rate of the immediately preceding response. The withdrawal of punishment will cause an increase in the rate of the response that terminates or avoids the aversive stimulus. Procedures which involve the use of aversive control have the advantage of being able to maintain behavior for a long period of time without experimenter involvement. In a successful avoidance procedure, for example, the animal's response resets the timer whose count-down would produce shock. The animal comes to respond regularly; his behavior is maintained by its success, i.e., no shock. If the experimenter then disconnects the shock apparatus and makes no other

changes, the behavior will typically continue for a long period of time.
Problems encountered with positive reinforcement such as deprivation and
satiation, are not encountered when aversive control techniques are used
over the short run.

1. Escape

Electric shock is administered at certain specified intervals for a
given duration of time, and the animal may terminate the shock by emitting
an appropriate response. For example, an operant chamber can be programmed to give a rat a 2-sec duration shock once every 20 sec; the shock
can be terminated by a lever press. This response is sometimes difficult
to shape. Rats have two clearly defined and mutually exclusive responses
to shock. Shock may either elicit running and an escape response, or the
shock can also elicit a freezing response in which the rat exhibits crouching,
along with sympathetic discharge, including piloerection, urination, and
defecation. The first set of responses makes it far more likely that the rat
will accidentally hit the lever during the running phase than under the latter
set of conditions. When the rat does accidentally hit the lever, the response
is reinforced by termination of shock, and therefore the lever press is
more likely to occur the next time shock is presented. When freezing occurs shock terminates after a specified interval and the freezing response
is also reinforced, even though this is not the "correct" escape response.
Unless this pattern can be changed such animals must be excluded from the
experiment.

2. Discrete Trial Avoidance Conditioning

In discrete trial avoidance conditioning a neutral stimulus is paired
with an aversive stimulus in the following manner:

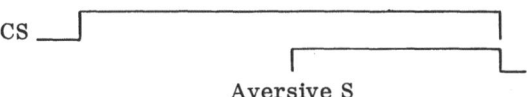

CS

Aversive S

The time interval between the onset of the neutral stimulus (conditioned
stimulus, CS) and the onset of the aversive stimulus (shock) is referred to
as the CS-UCS interval. A response which occurs during this interval has
the effect of terminating both the CS and preventing the impending shock
and is therefore termed an avoidance response. Often, experiments using
this paradigm are designed in such a way that a response (e.g., a lever
press) that occurs during the time period of simultaneous presentation of
the CS and UCS will also terminate the stimuli; this is essentially the escape

response described above. Generally, the CS-UCS interval is of the order of 10 sec, and the shock will have approximately a 5-sec duration.

There are a number of important parameters that must be carefully considered in order to obtain successful avoidance conditioning. Many of these apply equally well to the operant paradigm. The intensity of the shock can elicit freezing behavior which is incompatible with the development of avoidance behavior; on the other hand, shock levels that are too low will not serve as an effective aversive stimulus. Pulsating shock with certain minimal "off" times has been found to be more effective for the development of avoidance behavior than continuous shock. Other crucial variables include the duration of the CS, the spacing of trials, and handling of animals (13).

3. Continuous Avoidance

This paradigm is often referred to as nondiscriminated avoidance or Sidman avoidance (14). In this procedure, shock occurs at regular intervals in the absence of a lever pressing response; the interval between two shocks is referred to as an S-S interval (shock-shock) and may often be about 20 sec in duration. A response that occurs within the S-S interval has the effect of delaying the next shock. By pressing the lever at suitable intervals, then, shock can be postponed indefinitely. The delay of shock that occurs with each response is referred to as the R-S (response-shock) interval.

4. Factors Affecting the Acquisition and Maintenance of Avoidance

a. Rate of Responding as a Function of the R-S Interval. When the S-S interval is fixed at a certain value, the rate of responding decreases as the R-S interval increases; conversely, the R-S interval becomes shorter as the rate increases. However, the relationship between the rate of response and the R-S interval is not linear (Fig. 4). The R-S interval produces a maximum rate at a particular value; shortening or lengthening the R-S interval causes a decrease in the rate of responding. This may be understood in terms of the fact that with a long R-S interval, shock can be minimized even with a low response rate. In the case of the shorter R-S intervals, shock tends to occur after the response and therefore punish and suppress the rate of response (14).

b. Rate of responding as a Function of Shock Intensity. Some investigators have reported that the rate of responding on a nondiscriminated avoidance schedule is directly proportional to the shock intensity (15-19) while others have found that response rate decreases as a function of shock intensity (20, 21). Powell (22) by employing both single schedules of shock intensity (one shock for training) as well as multiple schedules (S^D for signalling the level of shock in effect) has been able to reconcile the differences between these experiments. His results showed that when rats were trained with one shock level, shifting the shock to a different level

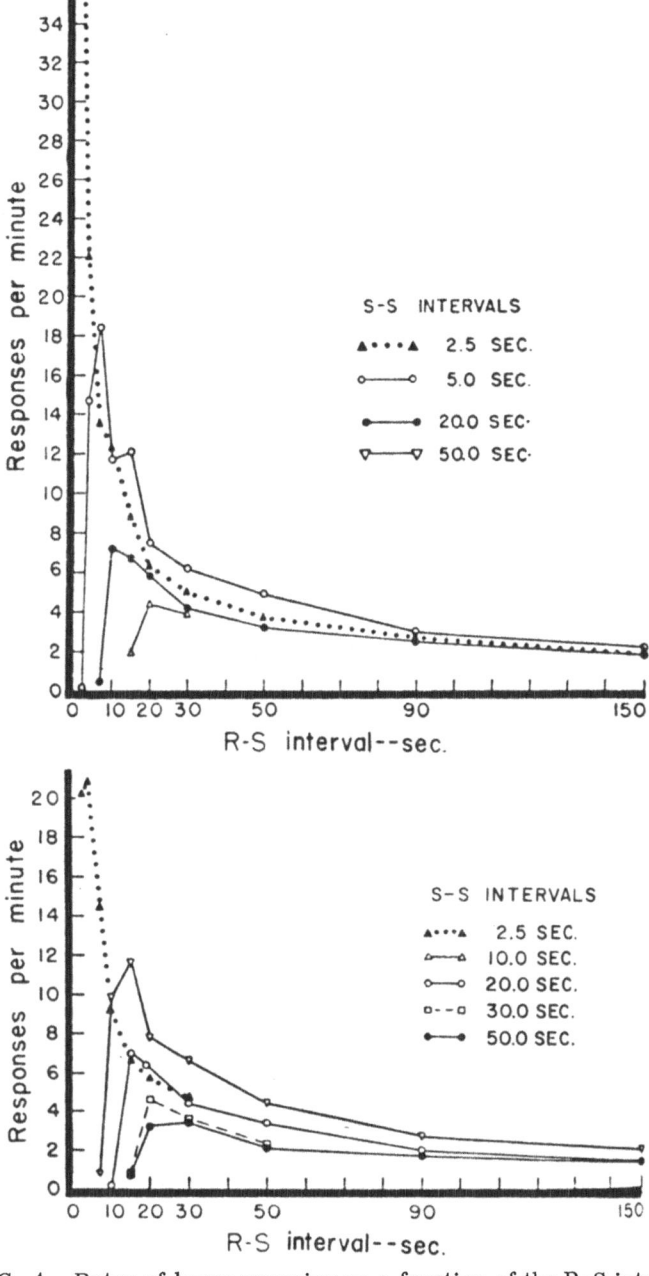

FIG. 4. Rates of lever pressing as a function of the R–S interval in a Sidman avoidance procedure. [From an article entitled "Maintenance of Avoidance Behavior" by Sidman in the Journal of Comparative Physiology and Psychology, 46, 253-261 (1953). Copyright 1953 by the American Psychological Association, and reproduced by permission.]

temporarily disrupted the avoidance behavior when the shock was either
increased or decreased. In the single schedule studies he found that rats
trained with the higher shock levels acquired the avoidance response more
consistently, and reached a higher terminal rate than rats trained with
lower shock intensities. Rats that were trained with a multiple schedule of
shock intensities had a higher response rate when the higher of the two
shocks was in effect.

   c. Shock as an Aversive Stimulus. In the vast majority of studies
employing aversive reinforcement, the reinforcer which maintains behavior is the termination or avoidance of electric shock. Shock is typically
administered through the feet of smaller animals (such as rats, mice, and
cats) or through implanted electrodes in the tail or foot of larger animals,
such as monkeys. Foot shock intensity can range from 0.1 to 5.0 mA.
The species involved, the duration of shock administration, and the conditions of temperature and relative humidity are factors which will be relevant determinants of the conductivity of electrical current from the grid
terminals to the surface area of the subject. Ohm's law $(I = V/R)$ is adequate for considering the relevant variables in this situation. R will be
determined by the internal resistance in the shock system plus the resistance of the animal, plus the resistance at the electrode-animal interface
which is highly variable. With low voltages (e.g., 100 V), variation in R
causes a manyfold variation in I, and since intensity is proportional to the
reinforcing and motivational aspects of the shock, such variations make experimental control difficult. One solution to this problem is through the
use of the constant current shocker. This type of system uses a high voltage (between 1000 and 2400 V) power supply. A relatively high resistance
must be placed in series with the animal in order to cut the amperage to
tolerable levels for the experimental animal. Under these circumstances,
the total resistance produced by the animal and the changes in resistance
produced by contact alterations are relatively negligible compared to the
large resistance already in the circuit and therefore the intensity of current
will be relatively constant.

   The animal must be in contact with opposite polarity grid bars in order
for the shock to be effective; if the grid bars have a constant polarity relative to one another, the animal will often develop an "avoidance response"
that will differ from the avoidance response defined by the experimenter.
To avoid constant relative polarity of the grid bars, a device called a grid
scrambler may be used. In order to adequately describe shock as a stimulus, experimenters must define the voltage, total resistance, and the duty
cycle of the grid scrambler.

   Kimble (16) reported that response latencies decreased as the intensity
of shock increased (over a range of 0.2 to 2.0 mA); latencies increased as

a negatively accelerated function of shock. In addition, it was found that the number of jump responses by the rat increased as shock increased, but that the number of flinch responses increased up to 0.3 mA but decreased at higher shock intensities. Moyer and Korn (20), on the other hand, reported that low shock produces more rapid acquisition of a shuttle box conditioned avoidance response than do higher shock parameters; they examined a range of shock intensities between 0.5 and 4.5 mA and found that that the optimal shock level was about 1.0 mA. The discrepancy between these two groups is not as great as would first appear since Kimble measured response over a lower range of shock intensities than Moyer and Korn. Moyer and Korn attribute the decrement in learning obtained at the higher shock levels to the possibility that a freezing response may have occurred more frequently in groups receiving higher shock than those with lower shock. To resolve these differences it is probably necessary to look at the detailed parameters of the experiment. For example, D'Amato et al. (23-25) found that discontinuous shock was more effective than continuous shock in facilitating discriminative avoidance learning of the bar press type. The pointed out that with discontinuous shock the rat is more likely to emit the escape response during the "off" period, so that with more "off" periods the escape response will be acquired rapidly; therefore, the avoidance response will also be acquired more rapidly.

E. Data Collection and Analysis

The most common instruments used for data collection are (a) digital counters and (b) cumulative recorders. For a digital counter, each lever press advances the counter; knowing the elapsed time, both average rate of responding and total output may be calculated from the counter. The cumulative recorder generates the type of record most frequently seen in the literature of operant behavior (Figs. 3 and 4). Records are generated by driving chart paper past a marking pen on an axis horizontal to the pen; each time a lever press occurs, the pen moves up by a preset amount; reinforcement can be indicated by diagonal movement of the pen. Other events, such as stimuli and "time-outs," can be marked by a second pen at the bottom of the chart. The slope of the line at any point in the cumulative record is proportional to the rate of responding. The advantage of the cumulative record over the counter is that is is possible to observe the pattern of responses engendered by a given schedule, periods of long pauses, and bursts of responding. The cumulative record contains every interresponse time coded as distance on the chart; with sufficient time the experimenter could tabulate all IRTs on one chart by hand measurements. Considering that there can be anywhere between several hundred to several thousand responses in a single session such hand measurements involve a prodigious amount of time.

The use of a digital computer greatly facilitates the collection and
analysis of interresponse time data because of the time and effort saved
in performing tabulations and calculations. It provides data for analysis
with a degree of sensitivity and complexity not possible with digital counters or cumulative recorders. On-line digital computer systems have been
described which not only measure the animal's performance, but also provide instant data analysis and subsequent feedback into the behavior system
(26). On-line systems have the advantage of being able to interact with the
animal's ongoing behavior and to modify the contingencies of reinforcement; however, on-line systems require that there be a one-to-one corre-
spondence between running time and computer availability, and this factor
can be a handicap to investigators sharing computer time.

Various off-line recording systems have also been described; these
systems require that the event be recorded as it occurs but allow for subsequent computer performance (27, 28). An off-line system in which data
from the session is recorded on an AM magnetic tape and transcribed into
IRTs at a later time has been described by Seiden et al. (29). In the latter
system each lever press is recorded as a spike with a duration of 5 μ sec.
Each reinforced lever press is coded as a double spike. Following the
recording session, the tape is played back into the computer at eight times
the recording speed, and each spike converts time on a clock within the
computer to a digital number and then reinitializes the clock. The digital
numbers so obtained are stored in the memory buffer until the session is
completed, and then they are written onto a magnetic tape which is com-
patible with an IBM 7094 computer. IRT histograms, continuous rate
graphs, numerical classifications of different IRT classes, as well as
statistical procedures are carried out on data obtained in this fashion.

III. ACTION OF DRUGS IN THE OPERANT PARADIGM

The effect of a drug on operant behavior can be measured as a change
in the rate of responding or a change in the typical distribution engendered
by a schedule. In the previous section we have discussed parameters of
the operant paradigm that have a functional relationship to the maintenance
of rate and distribution of responding (Fig. 1). Since these factors are
operative in the acquisition and maintenance of an operant response, it
follows that a drug might effectively modify one or more of these factors
and thereby modify behavior.

In the case of a positive reinforcement, the reinforcing value of the
stimulus (food or water) can be potentiated by prior deprivation. One of
the ways a drug can act is by altering the state of effective deprivation.
In this case, one would think of the drug as acting on PV (Fig. 1). In the
case of punishment, the aversive stimulus does not need to be potentiated

by antecedent manipulations; nevertheless, a drug could act by changing the animal's sensitivity to painful shock (as in the case of an analgesic). In many experiments that make use of operant procedures, animals are under the control of discriminative stimuli. These stimuli indicate when reinforcement is available and may also signal the schedule of reinforcement. In some instances, a breakdown of stimulus control may be indicative of drug interference with mediating sensory processes. A change in behavior may also be induced by changes in motor function.

It must be remembered that drugs have multiple actions, and knowledge of the pharmacological mechanisms of drug action has not been advanced sufficiently to allow one to predict where the action of a drug will be. Often, the behavioral actions of drugs are used as tools to investigate the physiology and biochemistry of certain neural systems involved in the mediation of behavior. It must be realized, however, that a given behavioral result can be explained in any one of several ways, and that no single experiment will have adequate controls to rule out all alternative interpretations that may account for the results.

Often attempts are made to classify drugs into general categories on the basis of their effects on behavior. Hence, drugs tend to be grouped as "depressants," "hypnotics," "tranquilizers," "stimulants," etc. For example, barbiturates are often classified as depressants while amphetamine type drugs are classified as stimulants. This classification holds in clinical work when the drug is prescribed in a particular behavioral context for a certain patient at an optimal dose. The classification of drugs as depressants and stimulants is by and large derived from their clinical utility and the ongoing behavioral status of the patient is assumed. However, drug action in experimental situations will vary; amphetamine which causes an increase in so-called "psychomotor" activity in human adults may cause a decrease in motor activity in hyperkinetic children. It has often been observed that certain barbiturates which depress activity and tend to induce sleep in most adults, paradoxically produce excitation in young children and older people. While under certain conditions these classifications may be descriptive of the actions of certain compounds on specific behaviors, this notion of classifying a drug without reference to the current ongoing behavioral repertoire or the dose of the drug is not consistent with even our relatively scant knowledge of the determinants of drugs' action on behavior.

A. Effects of Drugs on Behavior under the Control of Positive Reinforcement

The elegant experiments performed by Dews (30) illustrate that one must consider both the dose of the drug as well as the type of behavior in question when attempting to define the effects on behavior. Pigeons (31)

were trained to peck a key for a food reward. Two pigeons were trained on
an FR 50 schedule and two on an FI 15 schedule; after the effects of pentobarbital on the pigeons so trained were observed, the schedules were re-
served (FR pigeons switched to FI and vice versa) and the pigeons retested
after administration of pentobarbital. This experimental design made it
possible to compare the order of training as well as possible schedule interactions. The effects of pentobarbital on FI and FR behavior are illustrated
in Fig. 5. It is clear from these results that the effect of pentobarbital on
response output was dependent on the schedule of reinforcement. At a dose
of 1 mg of pentobarbital (per pigeon), FR rate was increased while the FI
rate was decreased. The fact that even the schedule-dependent aspect of
the drug-induced modification of behavior was dose dependent is indicated
by the 5.6 mg dose, at which point both the FI and FR rates are below the

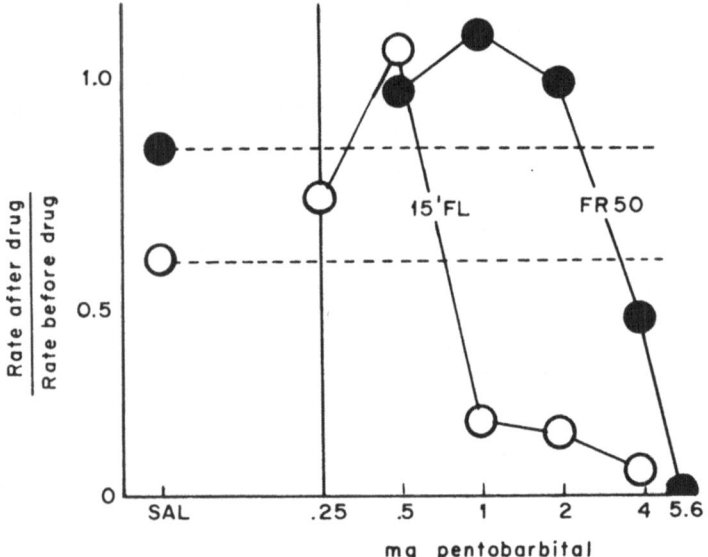

FIG. 5. Effect of pentobarbital on pecking behavior of pigeons. Log
dose-effect curves. Each point represents the arithmetic mean of the
ratios for the same four birds at each dosage level on each schedule. Open
circles: mean effects, birds working on 15-min FI. Solid circles: birds
working on FR 50. [Reproduced from an article by Peter B. Dews in the
Journal of Pharmacology and Experimental Therapeutics, 115, 393-401
(1955) with permission from the Williams and Wilking Company.]

levels of the control animals. This study provides an illustration of the necessity of examining several doses over different schedules. Had one dose on one schedule been examined, the conclusions might have been quite different than if another dose on another schedule was used.

The effects of schedule and dose-dependence of drugs are described in another study in which pigeons were trained on four different schedules of reinforcement, and subsequently injected with either methamphetamine, d-amphetamine, piperadol, pentobarbital, or scopolamine (32). Each schedule engenders the typical rate and pattern of responding as described in the previous section. The VI 1 schedule engendered a moderate response rate with an even distribution. The response rate on the FR 50 was high and the distribution of responding tended to be even. The FI 15 tended to generate a low response rate characterized by long pauses following reinforcement (Table 2). The FR 900 was characterized by a high rate of responding interrupted by long pauses sometimes occurring in the middle of the FR chain (FR straining). The results of this study showed that the effect of the drug was dependent on the schedule of reinforcement or perhaps the response rate engerdered by the schedule (Fig. 6). Methamphetamine caused an increase in response rate at several doses on the FI 15 and FR 900 schedule whereas the same drug over a similar dose range caused a decrease in the response rate on the VI 1 and FR 50 schedules. The latter two schedules contain IRTs in the control sessions that tended to be relatively short and contained few long IRTs, while the former two schedules contained both short and long IRTs. Dews suggested that the action of methamphetamine might be due to increased short IRTs and shortened long IRTs (32).

A similar drug-behavior interaction was obtained in a study by Schoenfeld and Seiden (33). It was found that α-methyl tyrosine (AMT) interferes with the maintenance of responding on both FR and FI schedules of reinforcement, but that an FR schedule was progressively affected as the size of the fixed ratio was increased; thus, on the FR the drug effect was a function of the FR parameters, i.e., the response output was decreased more on an FR 20 than on an FR 10 and more on an FR 10 than on an FR 5. In contrast the AMT effects on various FI schedules were not related to the length of the FI interval. Over the three time intervals tested, FI 30, FI 60, and FI 120 sec, the decrease in response output was the same for each FI schedule. IRT analysis of the schedules showed that the effect of AMT was to decrease the number of short IRTs and to increase the number of long IRTs. This occurred progressively over time; that is, at the beginning of the experimental session the distribution was the same as the placebo control, while at the end of the session there was a large difference (Fig. 7).

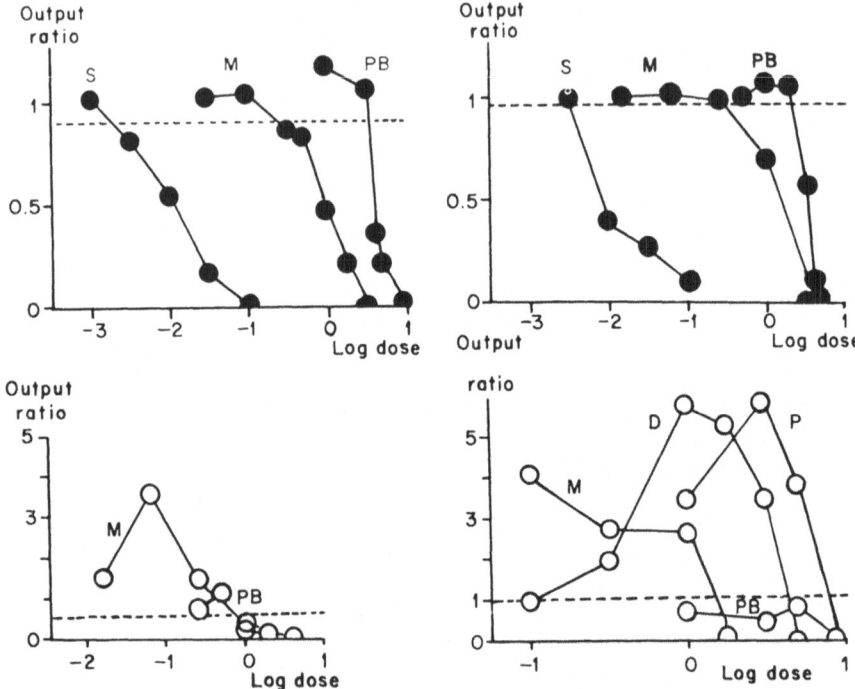

FIG. 6. Effect of drugs on performance on various schedules. Top
left--1-min VI. Top right--FR 50. Bottom left--15-min FI. Bottom
right--FR 900. (Output ratios calculated for whole experiments, not
just the first 15 min.) S--scopolamine. M--methamphetamine. PB--
pentobarbital. D--d-amphetamine. P--pipradol. The filled in circles
above and the open circles below are to emphasize the change in scale of
the ordinate between the upper pair and the lower pair. [Reproduced from
an article by Peter B. Dews in the Journal of Pharmacology and Experimental Therapeutics, 122, 137-147 (1958) with permission from the Williams
and Wilkins Company.]

Waller and Waller (34) have pointed out that the "schedule of reinforcement" encompasses at least two variables which can be specified independently. The contingencies of reinforcement engender a particular pattern
of responding which results in reinforcement (e.g., FI, FR, DRL, etc.)
and the density of reinforcement specifies the maximum rate at which reinforcement can be obtained assuming optimal responding. Variations in
the schedule and/or the density of reinforcement will cause a change in the
base line rate of responding; the studies cited above indicated that both can
be a variable in determining drug effects.

The result of these experiments indicate that the schedule of reinforcement and the dose of the drug are two critical determinants of a drug effect on behavior. In the experiment of Dews (Fig. 6), characterization of the effects of methamphetamine on the FI 1 min schedule would certainly differ from characterization of the effects based on the FR 900 schedule, and generalizations about the effects of methamphetamine based on a single schedule are likely to be incomplete. Similarly, these experiments show that whether a particular drug increases or decreases response output within a particular schedule, the response rate is very often a function of dose. Meaningful experimental design, then, must take account of these facts. At times, the cost in terms of time and money can be prohibitive in testing several different doses over several schedules of reinforcement; however, there does not seem to be a suitable alternative if one wishes to obtain a profile of the action of a drug on behavior. Failure to attend to these variables may be at least part of the reason that there are discrepancies among findings of investigators as to the characterization of the behavioral effects of drugs.

B. Effects of Drugs on Behavior under the Control of Aversive Reinforcement

Procedures employing the use of aversive control have been used for several reasons. In most studies in which aversive control is used to motivate behavior, the aversive stimulus is electric shock; therefore, it is not necessary to have a schedule of deprivation and be concerned with weight loss or various timing problems that can be involved with deprivation. From the standpoint of applicability to psychopathology, behavior produced under aversive control is often considered parallel to some of the behavior states seen in man that are treated with various behaviorally active drugs. It is very difficult to extrapolate from the behavioral paradigms used in laboratory animals to psychopathology in man in spite of the fact that some of the apparent variables that control the situation in the laboratory at first glance appear to be operative in man (see below). In addition, it is often difficult to apply pharmacological data across species. For these reasons, the experimenter must exercise caution in applying results obtained in a highly controlled laboratory situation to a clinical situation. For example, in some models of the conditioned avoidance response, the conditioned stimulus is said to produce a state of anxiety in the animal in the anticipation of pain, and it is because of reduction in this anxiety that the animal avoids shock. Effects of various drugs on avoidance and escape behavior have been recently reviewed (13, 35–37).

Certain drugs may affect the conditioned response while at the same time leave the unconditioned response relatively intact, whereas other drugs may affect both the conditioned and unconditioned response to the

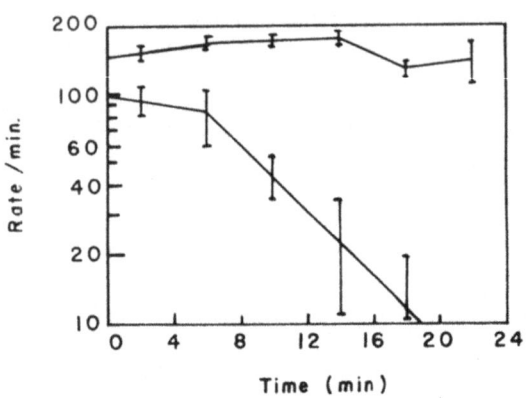

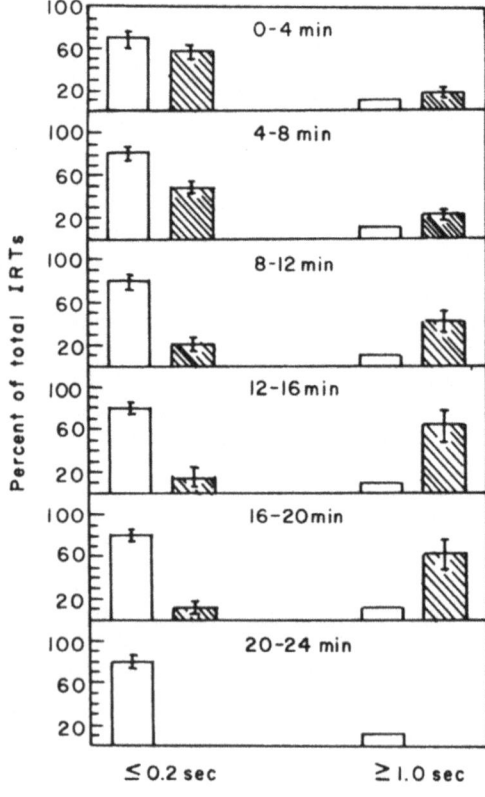

same extent at a given dose. In this section, some samples have been reviewed of behavior under aversive control which have been modified by drugs, in an effort to point out important variables to consider when using aversive control or attempting to compare aversively and appetitively reinforced behavior. A question posed by Kelleher and Morse (38) is whether the effect of a drug on a response that is under appetitive or aversive control can be attributed to the nature of the reinforcer (e.g., food vs. shock) or whether there is some other factor in the response pattern of distribution that is more salient in contributing to the drug effect.

1. The Use of the Conditioned Avoidance Response to Characterize the Effects of Drugs

The CAR has assumed importance for the study of two classes of drugs which overtly showed similar effects (13, 39), by virtue of the fact that it is able to discriminate between the effects of phenothiazine and barbiturate derivatives. It has been demonstrated by many investigators using different species that chlorpromazine (a phenothiazine tranquilizer) at certain doses inhibits the conditioned but not the unconditioned response. On the other hand, barbiturate derivatives have been found to interfere with avoidance responding only at doses which also interfere with escape responding (40).

Some of the earliest investigations, using a CAR as the baseline behavior, employed a pole-jump technique. The apparatus consisted of a metal grid floor with an insulated pole. An auditory stimulus preceded the delivery of shock to the rat. The escape response, or unconditioned

---

FIG. 7. Upper: Effect of α-methyl-tyrosine (AMT) (two doses of 75 mg/kg) on rate of lever pressing during FR 10 performance (8.0 hr after first dose). Each value represents mean rate per minute over consecutive 4-min periods of five animals. •, 30-min control session; X, 30-min test session.
Lower: Modal changes in IRT distribution produced by AMT. Each bar represents the mean percentage of total IRT's ≤ 0.2 sec or ≥ 1.0 sec for consecutive 4-min periods. The open bar represents 30-min test session; the hatched bar represents 30-min test session (two doses of 75 mg/kg of AMT). The mean values were obtained from IRT distribution graphs for the five rats used in the experiment described in the upper figure. Vertical lines represent standard error of the mean, *, standard error of the mean $\leq 1.0\%$. [Reproduced from an article by R. Schonfeld and L. Seiden in the Journal of Pharmacology and Experimental Therapeutics, 167, 319-327 (1969) with permission from the Williams and Wilkins Company.]

response (UR) consisted of jumping from the floor to the pole in response
to shock (UCS) and the avoidance response consisted of jumping from the
floor to the pole in response to the auditory stimulus (CS). Courvoisier et
al. (41) found that chlorpromazine could block the CR while the UR remained intact. Cook and Weidley (40) confirmed this finding and showed
that even at doses of chlorpromazine which block the CR by 100%, the UR
is only blocked by about 15%. Furthermore, they found that the selective
blockade of the response to the conditioned stimulus, although not unique
to chlorpromazine (morphine acts similarly), was not a general characteristic of all sedative drugs. Barbital and pentobarbital, drugs which are
often clinically used to induce sleep, showed similar effects on both the CR
and the UR and produced ataxia at higher doses (Fig. 8). This experiment
shows that chlorpromazine and morphine block the CR (avoidance) while
leaving the UR (escape) intact, and illustrates that the UR is a means to
evaluate the animal's ability to perform a motor task. If the rat can perform the motor task in response to the shock and there are no apparent
motor deficits, one may conclude that chlorpromazine blocks some component of the behavior system other than the _expression_ of the response.

It would depend on one's theoretical orientation as to which component
of the response chain one would postulate to be attenuated. For example,
if one considers avoidance behavior as being primarily motivated by fear
reduction, one would conclude that chlorpromazine acts by attenuating the
fear elicited by the CS, so that the motivation to perform the avoidance
response is reduced. If, on the other hand, one considers avoidance behavior to be maintained by the fact that a conditioned aversive stimulus is
terminated by the animal's response, then chlorpromazine could conceivably
act either through a sensory system, thereby reducing the detection of the
conditioned stimulus, or perhaps by reducing the memory of the connection
between the aversive shock and the conditioned stimulus. Regardless of
the theoretical orientation in explaining the action of a drug or in evaluating
the relevant determinants of a particular behavioral pattern, it is not possible from these CAR experiments alone to determine the mechanism of
action of chlorpromazine. Additional experiments which attempt to test
alternative explanations regarding the action of chlorpromazine are needed.
Almost without exception, any behavioral test system is so complex that a
given result may be explained by at least two or three mechanisms. The

FIG. 8. Upper: Block of conditioned response--chlorpromazine.
        Lower: Conditioned response. [Reproduced from an article by Cook
and Weidley in the Annals of the New York Academy of Sciences, 66, 740-
752 (1957), with permission from the New York Academy of Sciences.]

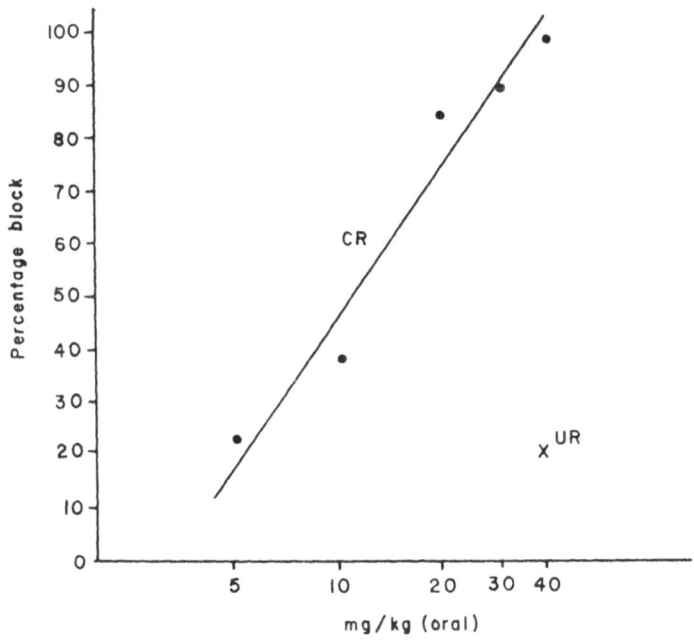

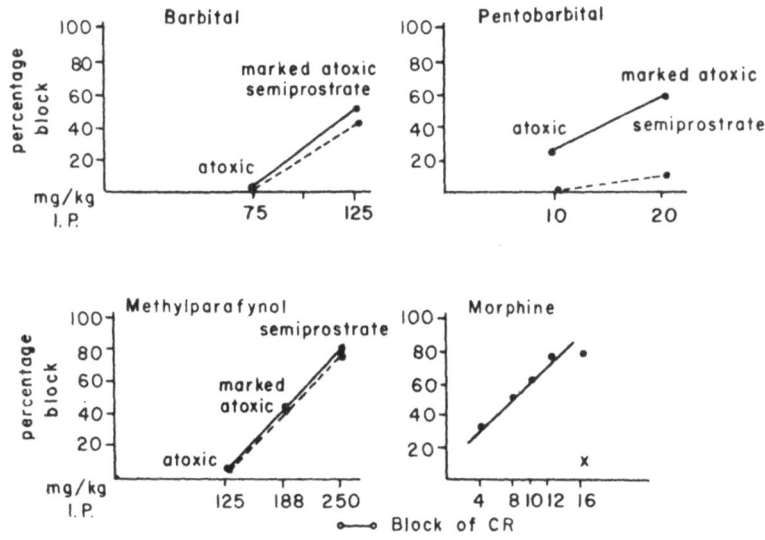

only way to arrive at a singular mechanism of action, if indeed one exists, is to test the drugs under new circumstances that test for these alternative explanations.

Miller et al. (42) performed an elegant experiment (43) using several experimental tests to determine whether the effects of chlorpromazine might be attributable to motor deficits and/or general sedative effects of the drug as opposed to effects on fear, arousal, or sensory systems. Using a two-way shuttle box with a five-second buzzer as a CS, they established that in this situation chlorpromazine produced a dose-dependent block of the CR. They reasoned, however, that a decrement in CR could be indicative of either motor deficits or decreased fear elicited by the CS. If one could find a drug that produced motor deficits similar to chlorpromazine but no deficits in the performance of the CR, then by a process of elimination one could conclude that systems other than motor must be involved in chlorpromazine action. Accordingly, a shock escape runway was designed in which the latency of escape over an electrified grid was measured. They were then able to quantitate motor deficits (in terms of time required to transverse the runway) produced by an optimal dose (in terms of two-way avoidance); it was possible to find a dose of sodium phenobarbital which produced the same slowing of the escape response as chlorpromazine and therefore, presumably, the same motor deficits. If chlorpromazine was acting only on the motor system then one would expect that the behavior pattern after sodium phenobarbital would be similar to the pattern observed with chlorpromazine and one would expect that phenobarbital would interfere with the CR at this dose. Phenobarbital did not, however, interfere with the CR, and on this basis the authors concluded that chlorpromazine did not exert its primary action on motor systems. These data present evidence that the action of chlorpromazine is similar to the action of phenobarbital in one test system, but can be differentiated from phenobarbital in another test system.

Cook and Catania (35) have shown that the magnitude of a drug effect in a CAR paradigm is in part dependent on the nature of the CS. In this experiment, dogs were trained to avoid shock; one CS was a tone, the other an injection of epinephrine. It was shown that chlorpromazine interfered more with the response mediated by the epinephrine CS than the tone CS. Thus it is of importance that the experimenter specify the nature and intensity of the CS.

a. Interaction between Stimuli Parameters and the Maintenance of the CAR. Certain qualitative and quantitative aspects of the conditioned stimulus were found to be of importance in producing learning of an avoidance response in either a shuttle box or an operant situation. Several investigators have shown that variations in the parameters of avoidance conditioning lead to differences in the observed drug effect. For example, Low et al. (44) trained rats to perform avoidance in a shuttle box situation and used three different CS/UCS intervals (5, 10, or 15 sec). The drug did not produce the dose-dependent depressions of conditioned responses with a 5- and 10-sec CS/UCS interval; with the 15-sec CS/UCS interval it did (45-47). It has been noted that a buzzer (mixed frequencies) is superior to a specific tone (one frequency) in serving as a CS in a shuttle box; experiments have been performed which indicate that it is not simply a function of the intensity of the buzzer over the tone that makes the buzzer a superior conditioned stimulus (48). Furthermore, the onset of a buzzer CS is more effective than termination of the buzzer and this may be due to the fact that onset of the buzzer produces a startle response (49). In a shuttle box avoidance, the speed of acquisition decreases as the intertrial interval increases; however, as the intertrial interval is increased over 5 min the rate of acquisition also decreases (50). On the other hand, in a discriminated avoidance response in an operant situation, as the intertrial intervals increase the rate of responding decreases (51, 52). Avoidance learning can also be a function of different training conditions as well as strain differences (53).

It is clear that shock level will affect the rate of acquisition of a shock-reinforced response. A drug which produces hyperalgesia or analgesia could conceivably modify avoidance behavior through this mechanism. The question of sensitivity to pain is relevant to the question of drug-induced modification of behavior maintained by aversive reinforcement. Tenen (54) has shown that p-chlorophenylalanine (PCPA) can increase the jump threshold to electrical shock in a flinch-jump apparatus. PCPA treatment facilitated the acquisition of a CAR, an effect which was apparent when the animals were trained under low shock values. However, if the intensity of shock was raised differences between the control rats and those treated with PCPA were eliminated. This clearly illustrates that shock intensity is a relevant variable in the acquisition of the CAR.

   b. Measurement of Sensitivity to Electrical Shock. The "flinch-jump" is used to measure the sensitivity of an animal to electrical shock (54-56). A rat is placed on a grid and by increasing and decreasing the amperage across the grid, the threshold at which the rat flinches and the threshold for a jump to occur can be determined. A flinch is defined as a noticeable convulsive-like movement (crouch or startle) in response to the shock, and a jump is defined as a movement in which the rear paws leave the grid in response to the shock. Since this procedure requires experimenter observation, it can be somewhat subjective. For this reason, Weiss and Laties (57) have introduced a shock sensitivity measuring procedure making use of operant techniques; this procedure is called a "titration schedule." Rats are placed into an operant chamber with a grid floor; at regular intervals (called the i-i interval) the intensity of the shock is increased by a fixed amount. If the rat presses the lever the intensity of the shock is decreased by a fixed amount; it has been observed that rats can maintain the shock at a constant level by lever pressing. Weiss and Laties (58), by adjusting the rate of change of the shock, have carefully explored a range of variables which determine the rate of pressing. In general, it is found that as the i-i interval is shortened the titrated level of shock intensity increases.

This method has been used to test the properties of a number of analgesics. Administration of the "known analgesics," such as morphine and aspirin, result in the rat titrating the shock to a higher level. Control experiments in which phenobarbital was used indicated that the effect was not produced by general depression of activity, for although there was suppression of motor activity to the point of ataxia with phenobarbital, there was not a significant increase in the median shock level. On the other hand, the fact that the salicylates could disrupt a DRL or a VI schedule under the control of positive reinforcement indicates that these drugs may owe their effect on shock titration in part to some disruption of timing behavior. The experiments of Weiss and Laties (59) illustrate the necessity of running collateral experiments in any attempt to isolate a single variable in a behavioral situation. The number and type of collateral experiments that are necessary depend on the number of alternative explanations that can account for the data obtained in a given experimental situation.

2. The Conditioned "Emotional" Response (CER)

If one makes certain assumptions about the behavioral implications of the state of anxiety, the CER may be considered as an experimental analog of anxiety or fear. The basis for the comparison rests on the assumption that "fear or anxiety" can interfere with ongoing behavior. In the CER paradigm, behavior is brought under the control of a positively reinforcing stimulus. When lever pressing for positive reinforcement is stabilized a neutral stimulus (CS) is then presented; the termination of the neutral stimulus is paired with an unavoidable shock (UCS). Repeated pairings of the CS and shock for three to seven trials results in a discriminated conditioned suppression of lever pressing. Suppression begins with onset of the CS and ends upon termination of the CS. Suppression of lever pressing also occurs early in the conditioning procedure in response to the shock alone, but the shock-induced suppression is usually transient. When the animal begins to show suppression of the lever pressing in response to the CS, suppression of responding during the post-shock interval is minimal. There are several ways to measure the strength of the CER; the most common makes use of the idea that the CS-shock pairing evokes a response that "competes" with the ongoing operant. A convenient measure of this is the relative rate of response before (A) and during the CS (B). The inflection ratio $(B-A)/A$ (60) and the ratio described by $B/(A+B)$ (61) normalize data so that different animals may be compared independently of the rate of lever pressing.

Tenen (62) has employed the time it takes for the animal to resume lever pressing after shock (recovery time) as a measure of CER strength.

The CER paradigm was initially developed by Estes and Skinner (63) and has been used to test drugs in the "tranquilizer" class, i.e., drugs which may selectively reduce responses to aversive stimulation (43). Although the literature has not been completely consistent, the CER elicits not only suppression of ongoing behavior, but a variety of autonomic changes: changes in heart rate, defecation, pupillary dilatation, urination, and piloerection. The autonomic changes are accompanied by changes in postural adjustments that are characterized as "emotional," such as crouching, wild running, and jumping or freezing. Since it is tempting to draw the analogy between human behavior that is disrupted by certain environmental contingencies that have an aversive connotation and the CER, this paradigm would seem ideal for measuring the effects of drugs which would be potentially useful in clinical medicine.

A review of the behavioral literature reveals that despite the apparent utility of the CER for the screening of drugs which might potentially reduce aversive control of behavior, some investigators report that certain drugs such as reserpine and chlorpromazine attenuate a previously established CER (64, 65), while other investigators have found that these and additional compounds which are clinically useful do not (66). Experimental analysis has revealed several important parameters of the CER paradigm that have profound effects on the development and maintenance of a CER. Perhaps closer analysis of the behavioral situation where conflicting result are obtained would help to explain the differences. For example, there is evidence that the intensity of the UCS can influence the effects of a drug in attenuating a CER. Appel (65) has shown that reserpine attenuated a CER established with low but not high intensity shock.

Drug effects have been related to the nature of the CS (67). The interaction between the respondent conditioned suppression (CER) and the strength of the ongoing competing operant behavior is also an important variable. When sweetened milk was compared with water as a base-line reinforcer, it was found that both acquisition and extinction of a CER were slower with sweetened milk (68). Stein et al. (69) showed that animals trained on a CER exhibited suppression of the operant behavior if and only if there was a chance to "make up" reinforcements lost as a result of the suppression. Therefore, the suppression of the operant behavior is related both to the nature and the scheduling of the reinforcer. Harvey et al. (70), in an elegant study, showed that rats with lesions in the septal area of the limbic system showed an attenuated conditioned emotional response, where the lever pressing response was reinforced with water. If these rats were given access to water prior to the test session, their expression of the CER was the same as that of a sham operated group; this indicates that the extent of deprivation (see Fig. 1, PV) can play an important role in the expression of a CER and that this factor would need to be considered in measuring the effect of a drug-CER interaction.

Blackman (71) has presented intriguing data which indicate that the expression of the CER may be dependent on the rate of lever pressing. In this study, rats were trained on a VI one minute and a CER was imposed on the operant. The rate was lowered or raised in different rats by appropriate pacing procedures, and the CER was retested. The rat with the high rate retained the CER while the rat with the low rate showed complete loss of the CER. When the pacing procedures were reversed so that the rates of the two subjects were reversed, the CER in the rat with the lower rate was again lost but that in the rat with the high rate was still apparent. Since chlorpromazine and reserpine, as well as many drugs which attenuate the CER, lower response rate the question arises as to whether the behavior is controlled by fear, or whether the attenuation is simply a function of the rate of lever pressing.

C. Reinforcement-Dependent Drug Effects

It is sometimes assumed that the clinical efficacy of tranquilizing drugs is reduced if they suppress behavior maintained by both positive and negative reinforcement. It is of interest to determine if drugs selectively affect responding maintained by a particular reinforcer, and to what extent simple variation of deprivation or shock intensity will affect the baseline behavior and the response to a drug. This is not a simple question to answer because changes in both schedule and potentiating variables which change the potency of stimuli as reinforcers will also change the baseline rate of responding. Therefore, it is very difficult to differentiate between a change in PV or rate change as the causative factor in the effect of a drug.

Weissman (72) trained rats to press a lever for food and shock avoidance on a multiple schedule (extinction, CRF, RS10, SS10). A clicker indicated that a CRF schedule which was reinforced with food was in force; a tone indicated that S-delta (no reinforcement available) was in force, presence of a houselight indicated that a continuous avoidance procedure was in force with a shock-shock interval of 10 sec (SS 10) and the response-shock interval of 10 sec (RS 10). Control responding on this program consisted of a fairly rapid response rate during the avoidance component of the schedule, a slower rate during the CRF for food, and cessation of responding during S-delta. Several drugs were tested in this experiment and it was found that there were differential effects of amphetamine, chlorpromazine, morphine, and iproniazid on the positively and negatively reinforced components of the schedule. However, the baseline rates of responding on the positive and negative schedules were different, and if the differential drug-response is to be attributed to the reinforcer, as such, then the difference in baseline rate must be taken into account. In a similar study, Cook and Catania (35) found that chlordiazepoxide interfered with a shock-avoidance schedule while leaving an FR schedule for food reinforcement essentially intact on a concurrent FR-avoidance schedule. However, as in the Weissman study (72) the response rate for the two reinforcers was not the same, and as the authors point out one cannot differentiate the effects of the reinforcer in the face of differing response rates. Ray (73) also reports selective blockade of an avoidance response using a discrete trial procedure. The latency of the response rather than the response rate after presentation of the discriminative stimulus was measured. Using this paradigm avoidance latencies were increased by meprobamate, chlorpromazine, and reserpine at doses which have no effect on appetitive latencies. It is difficult from the data presented to ascertain if the pre-drug latencies are equal with both reinforcers. Moreover, it would seem desirable to know if the particular shock and deprivation levels used in these studies could differentially affect not only the baseline latencies but also the response to drugs. In this particular experiment, 22-26 hr of deprivation, and 1.0 mA of shock were used. If one is to claim that the shock procedure is more sensitive to the disrupting effect of the drug, then one also assume that the shock and deprivation levels used are equivalent with regard to their ability to maintain behavior. Unless there is a reason for the use of a given parameter of shock, deprivation, or amount of reinforcer, then it is necessary to use more than one value. If shock control of ongoing behavior is weak, then it would seem apparent that it may be disrupted before appetitive control and vice-versa.

Kelleher and Morse (38) trained monkeys to either avoid shock or be reinforced with food on both FR and FI schedules. Deprivation conditions and shock levels were arranged such that the rate of responding on the FI schedule was independent of the reinforcer, and the same held true of the FR schedule. When the monkeys so trained were given chlorpromazine and amphetamine the effects of these drugs were clearly dependent on the schedule of reinforcement rather than the nature of the reinforcer (Figs. 9 and 10). In this study, both reinforcers within either schedule produced a similar rate; therefore, it appears that the effect of the drug is more dependent on the baseline rate of responding than on the nature of the reinforcer. These results are consistent with those obtained by Waller and Waller (34).

The work done in these two studies considers only a limited number of schedules, drugs, shock values, and deprivation conditions probably does not allow one to generalize that all drugs effect are rate-dependent. There is little in the literature that I have seen that contradicts the work

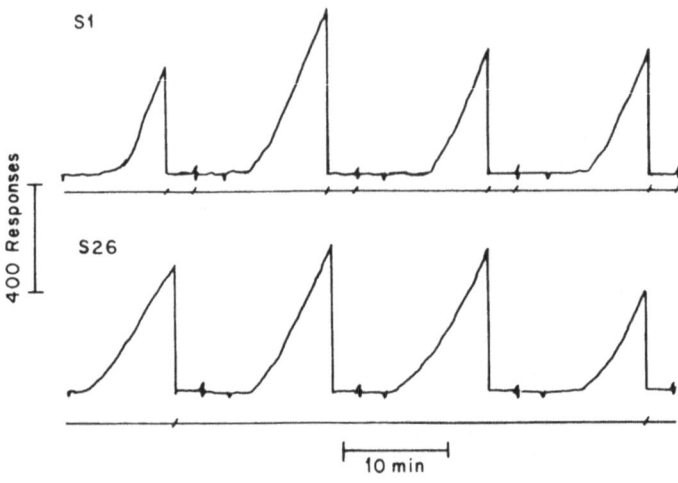

FIG. 9. Performance maintained by a positive reinforcer (food) compared to performance maintained by a negative reinforcer (electric shock)
under multiple FI, FR schedules in the squirrel monkey. Ordinate, cumulative number of responses; abscissa, time. Upper records: typical performance of monkey S1 on the multiple schedule of positive reinforcement.
At the beginning of the records, the 10-min FI schedule was in effect in
the presence of a white stimulus. At reinforcement, the cumulative recording pen resets to the bottom of the record, and a short diagonal stroke
appears on the event record. Following reinforcement, a pattern of horizontal bars was presented on the stimulus panel for 2.5 min; in the presence of this pattern, responses had no specific consequences and reinforcements never occurred. The next short diagonal stroke on the cumulative
record indicates that the 30-response FR schedule was in effect in the
presence of a red stimulus. Again, the cumulative recording pen reset to
the bottom of the record at reinforcement, a short diagonal stroke appears
on the event record, and the pattern of horizontal bars was presented for
2.5 min. This cycle was repeated throughout each session. Lower records: a sample of the performance of monkey S26 on the multiple schedule
of negative reinforcement. The sequence is visual stimuli and corresponding schedules is the same as in the upper records. The cumulative recording pen reset to the bottom of the record when a response terminated the
white or red stimulus light. The short diagonal strokes on the event record indicate shock deliveries. Note the similarities of the patterns of
responding under these multiple schedules of positive and negative reinforcement. [Reproduced from an article by R. Kelleher and W. Morse in
Federation Proceedings, 23 (4), 808-817 (1964) with permission from the
Federation of American Societies for Experimental Biology.]

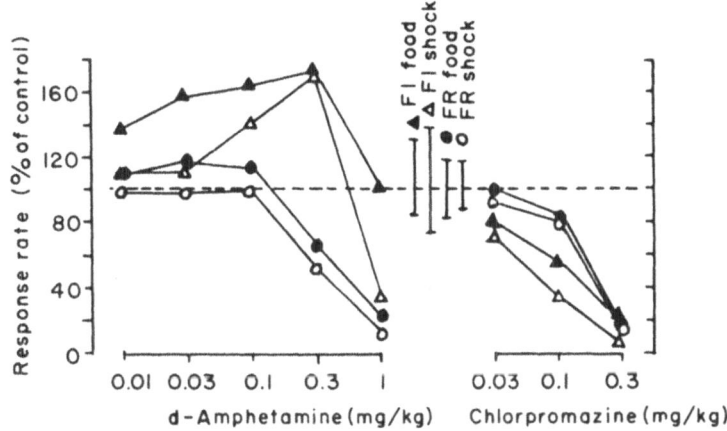

FIG. 10. Effects of d-amphetamine sulfate and chlorpromazine hydrochloride on rates of responding under multiple FI, FR schedules of positive and negative reinforcement. Three monkeys were studied on each multiple schedule. Each drug was given intramuscularly immediately before the beginning of a 2 1/2-hr session. At least duplicate observations were made in each monkey at each dose level. Summary dose-effect curves for the four component schedules were obtained by computing the means of the percentage changes in average response rates from control to drug sessions. The dashed line at 100% indicates the mean control level for each component. The vertical lines in the middle of the figure indicate the ranges of control observations, including saline injections, expressed as a percentage of the mean control value. Note the general similarity of the pairs of dose-effect curves for fixed-interval and for fixed-ratio components. [Reproduced from an article by R. Kelleher and W. Morse in Federation Proceedings, 23 (4), 808-817 (1964) with permission from the Federation of American Societies for Experimental Biology.]

of Kelleher and Morse; on the other hand, experimenters have not always been concerned with rate as a crucial variable and have therefore not always controlled for rate as a variable in their experiments.

1. Drugs as Reinforcers

In the main, some of the most apparent ways drugs can act on behavior that is under the control of reinforcing stimuli have been discussed. It is quite apparent that at least certain drugs (e.g., morphine, cocaine, amphetamine, and the barbiturates) will, under some conditions, act as reinforcers. This is evidenced by the fact that primates (including man) as well as rats will self-administer these compounds when given the opportunity to do so. The various techniques available for the study of drug self-administration goes beyond the scope of the chapter [for an excellent review of this literature see Thompson and Schuster (74)]. Briefly, in the paradigm for drug self-administration in animals, a catheter is implanted in a vein or artery. The reinforcement for a lever pressing response is the delivery of a drug, such as morphine, into the blood stream. Drugs that are self-administered bear a striking resemblance to other appetitive reinforcers such as food and water insofar as they can maintain behavior on a variety of schedules of reinforcement, and when they are withheld, the behavior which they reinforced will eventually extinguish. However, drug deprivation does not have meaning in animals without prior experience with the drug. Typically then, drug dependence must be established before a compound such as morphine can become a reinforcer. The amount of drug needed to establish dependence to the point that withdrawal of the drug is a sufficient antecedent manipulation to motivate working for the drug will vary with different classes of compounds, the dose of the drug, and the species. It should also be pointed out that drugs as reinforcers are unique in that a sufficiently large dose of the "reinforcer" can be lethal to the animal. However, even with certain problems these techniques of drug self-administration have been used to effectively study the physiological, pharmacological, and environmental variables that influence this behavior.

2. The Neurochemical Basis of Behavior—Methods of Study

Evidence derived from drug action on the autonomic nervous system indicates that most drugs act by altering the metabolism, uptake, storage, or retention of compounds involved in synaptic transmission. Although the particular transmitters in the central nervous system have not been completely delineated, there is substantial evidence that there are at least four, and probably more, compounds which can be classified as transmitters: acetylcholine, norepinephrine, dopamine, and serotonin. Much of the evidence for the transmitter function of these compounds derives from studies which have demonstrated that drugs which alter electrophysiological function in the brain also alter some aspects of the metabolism and utilization of these compounds. In addition, these compounds have been shown to exist in nerve endings in the brain; the enzymes for their synthesis and degradation are localized in the CNS in such a place as to be consistent with their role as neural transmitters. In some instances it has been demonstrated that stimulation of certain nerve fibers causes the release of the compound, and topical application causes changes in firing rate of neurons.

It has also been demonstrated that many behaviorally active drugs produce changes in the storage, release, and uptake of the synaptic transmitters. The drug-induced biochemical changes may tell us something not only about the biochemical mechanisms by which the drug exerts its effect on behavior, but may also yield information as to the biochemical processes in the brain (vis-a-vis either synaptic transmission or other metabolic processes) that are necessary for the development and maintenance of behavior. Since it is well known from work in neuropharmacology that drugs can alter various aspects of CNS biochemistry, drugs have often been used as a tool to manipulate the biochemistry and thereby study the role of the putative transmitters in behavior.

Causal relations between biochemical events in the central nervous system and behavior are difficult to establish. If there are several drugs which produce the same biochemical effects on the CNS as well as the same effect on the behavior of the organism, then one would be tempted to draw a causal relationship between the chemical and behavioral changes. This ideal situation of finding several different drugs with similar chemical and behavioral effects is seldom obtained, and furthermore, the approach has some difficulty even when it occurs. This can best be illustrated by remembering that drugs rarely act on a single system within the organism, but probably act on several chemical systems; the drug effect could conceivably be due to an effect on the biochemistry of one or more of these systems, and to assume that the effect is attributable to the change in the particular system that is being measured is a weak assumption.

If several drugs have the same biochemical and behavioral effects this would constitute evidence that the drugs act via this particular biochemical mechanism to produce the behavioral effect, but a common effect on some other biochemical system cannot be ruled out completely. Further studies including the use of drugs that can act as antagonists or blockers of the chemical effect in question help to narrow down the possibilities. Certain of the drugs that are typically considered in this regard act by depleting neurotransmitters or interfering with their synthesis. In this event it is possible to replete the transmitters with the use of appropriate chemical precursors. If the behavioral effect of the drug can be reversed by the use of appropriate drug antagonists or by repleting the lost transmitter substance, this strengthens the argument that the behavioral effects of the drug are mediated through the transmitter. The particular antagonist and chemical precursor employed would depend on the chemical system involved. (See Refs. 75-77.)

3. Drug Dosage

We have discussed in preceding sections how the effect of a drug is a complex function of both the ongoing behavior of the animal and the dose of the drug. While we have discussed the former (ongoing behavior), little has been mentioned regarding the factors involved in the administration of drugs over an effective dose range. We must start this discussion with the realization that since all drugs are toxic when given in high enough doses, the degrees of freedom in selecting a dose of the drug will be limited by this fact. The toxicity of a drug is often measured in terms of its lethal effects. The LD_{50} (lethal dose) is defined as the dose of drug that will kill 50% of a population of animals. On the other hand, the ED_{50} (effective dose) is defined as the dose that has an effect on a certain measure in 50% of a population. Both of these figures are obtained by using a large dose range (between 10 to 100-fold). The drug effect, as measured by the LD_{50} or the ED_{50}, is a function of several important variables as discussed below.

4. Route of Administration

There are at least five routes by which a drug can be introduced into an organism: orally (p. o.), intraperitoneally (i. p.), intramuscularly (i. m.), intravenously (i. v.), and intracisternally or intraventricularly. Each route has its own special advantages and disadvantages, but the major point to emphasize is that route will make a difference in the drug effect.

5. Drug Vehicle

Most drugs must be dissolved or suspended in a liquid before administration. While the nature of the carrier vehicle will to some extent be determined by the route of administration, other factors such as the solubility of the drug in the vehicle also enter into consideration.

6. Strain and Species Difference

It is well documented that strain and to a much greater extent, species differences, have different sensitivities to drugs. The investigator must proceed with caution in extrapolating dose response information from one type of organism to another.

IV. SUMMARY

In this short chapter, we have seen several experimental variables that can affect both the baseline behavior and the manner in which that behavior may be modified by a drug. Some of the parameters may be varied in a linear fashion (e. g., shock intensity, degree of deprivation, amount of reinforcement) but others add even greater complexity to the situation in that they may be varied along several dimensions which may not be completely scaler (e.g., schedule of reinforcement, prior handling, strain of animal, etc.). It is very often difficult to decide on the exact parameter for any given variable, and is is somewhat staggering to think about the amount of experimental work it would take to systematically study all of the variables (see Table 3). It brings to mind Carroll's poem "The Walrus and the Carpenter:"

The Walrus and the Carpenter
Were walking close at hand;
They wept like anything to see
Such quantities of sand;
"If this were only cleared away,"
They said "it would be grand!"

"If seven maids with seven mops
Swept it for half a year,
Do you suppose," the Walrus said,
"That they could get it clear?"
"I doubt it," said the Carpenter,
And shed a bitter tear.

TABLE 3

Basic Variables in Experimental Procedures

General variables

Age	(17, 78-80)
Species/Strain	(81-83)
Rearing/Housing	(84-102)
Stimulus Control	(103-118)
Training	(117, 120)

Appetitive variables

Deprivation	(15, 121-123)
Reinforcement	(15, 125-137)
Schedule of Reinforcement	(30-32, 137-138)

Aversive variables

Shock	(16-22, 24, 25, 139, 140, 142)

Nevertheless, the sand must be removed from the shore if we are to understand and be able to characterize the actions of drugs on conditioned behavior. From a collective point of view it would be enormously helpful if various bits of data from one laboratory could somehow be compared and coupled with data from other laboratories. This is virtually impossible if the variables are not adequately defined, and extraordinarily difficult if they are defined but not consistent from laboratory to laboratory. While there is no suitable alternative to the collection of large amounts of data, it does seem that a systematic and reasoned approach both to the design and analysis of experimental drug-behavior studies would be facilitated by the development of theoretical positions. The past three decades have witnessed the growth of a number of weak theories and for this reason, many experimental phychologists are today "anti-theoretical." While I favor the rejection of weak theories, i.e., those that are not testable and do not stimulate research resulting in their own modification, I see no reason to be anti-theoretical in principle. The growth of physics, chemistry, and biology has been stimulated and built on theoretical positions. It seems to me that it will become impossible to formulate new experiments or transmit information coherently unless a workable conceptual framework is developed by the experimental scientists working in this field.

APPENDIX I

A. Detailed Protocol for an Experiment

It is difficult to generalize about the precise design of a protocol regarding the dose of a drug, the number of animals included, or the particular schedule employed. The particular design depends on a variety of factors including the drug, the schedule of reinforcement, the species, and the type of measurements sought. The following protocol sheets give an example of an experiment performed in this laboratory (Roger M. Brown, Ph. D. dissertation). In this experiment Dr. Brown was interested in measuring the effects of d-amphetamine, water-loading, and their interactions on IRT distributions using a VI 20-sec schedule of reinforcement. This protocol is presented as a representative model with the view of helping a novice set up a new experiment. In this particular experiment n = 8.

PROTOCOL
_____

From

Day 1: Arrival of 8 male Sprague-Dawley rats (Holtzman line). Rats are removed from cartons, have their ears punched with ID numbers, and are weighed. Housed 2/cage with food (rat chow) and water ad lib.

Days 2-4: Rats "gentled" for one to two minutes daily. Weighed. Food and water ad lib. Gentling procedure consists in holding the rat (preferably close to the body of the experimenter and firmly supported) for a minute or two. We have found that doing this before conditioning makes the rats easier to handle and condition. Over the first four days with food and water ad lib, a weight gain of between 5 to 20 g can be expected.

Day 5: Pull water bottles from cages and begin water deprivation. Continue daily gentling procedure and weighing.

Day 6: Water for 5-10 min. Rats are now watered in a separate cage (between 5-10 min) individually so water consumption may be measured.

Days 7-9: Continue as day 6. Weights should stay stable or perhaps show a small gain. Animals usually drink about 10 ml of water per day to maintain weight gain.

Day 10: Shaping of lever pressing response in operant chamber (see text). This should take about 15 min per rat.

Days 11-12: Pressing CRF schedule for 15 min. In this experiment all animals were given supplemental water (7 min free access) 10 min after the session. All rats weighed to observe possible weight loss from day to day.

Day 13 to stability: VI 20 sec. Stability defined as not more than 10% variation from the mean of 3 previous days.

Begin experimental day 1 as below. Note that this experimental design varies treatments for each animal.

B. Effect of Satiation (Water Loading), Water Deprivation, and d-Amphetamine on Lever Pressing Behavior under a VI 20-sec Schedule of Water Reinforcement

Each animal will receive the following treatment:

Water loading by allowing 2.5, 5, 10, and 20 min of free water time immediately before the 30-min session.

Deprivation: In addition to a normal 24-hr deprivation, animals will be deprived of water for 48 and 72 hr. This is not strictly 48 and 72 hr of deprivation as animals will be given their daily 30-min session with reinforcement available. The free water usually given 10 min post session will be withheld for 48 and 72 hr.

<u>d-Amphetamine</u>: Doses of 0.25, 0.5, 1.5, and 0.75 mg/kg of d-amphetamine sulfate will be given i.p., 30 min prior to the session.

<u>72-hr deprivation + amphetamine</u>: To test for deprivation-amphetamine interaction, animals will receive the four doses of amphetamine after withholding post-session free water for 72 hr.

<u>Water loading + amphetamine</u>: Animals will be given the four doses of amphetamine and in addition will be allowed a period of free water immediately prior to the session (tentatively, 2.5 min of free water will be used).

Experiment IA: Involves data from amphetamine, deprivation, and water loading treatments.

Experiment IB: Involves the interaction data.

Procedure: (Data will be taped during the days marked with an "*." Control days will be signified by "C.")

	Day	1 & 2	3 & 4	5 & 6	7 & 8
IA	1*	c	c	c	c
	2*	saline	c	c	c
	3*	0.25 d-A	20 min	2.5 min	20 min
	4	c	c	c	c
	5	c	c	c	c
	6*	c	saline	c	saline
	7*	2.5 min	1.5 d-A	5 min	0.25 d-A
	8	c	c	c	c
	9*	c	c	saline	c
	10*	5 min	10 min	0.26 d-A	10 min
	11	c	c	c	c
	12	c	c	c	c
	13*	saline	c	c	c
	14*	0.5 d-A	5 min	10 min	5 min
	15	c	c	c	c
	16*	c	c	c	c
	17*	10 min	saline	20 min	saline
	18*	c	0.75 d-A	c	0.5 d-A
	19	c	c	c	c
	20*	c	c	c	c

	Day	1 & 2	3 & 4	5 & 6	7 & 8
	21*	48 hr dep	48 hr dep	48 hr dep	48 hr dep
	22*	72 hr dep	72 hr dep	72 hr dep	72 hr dep
	23	c	c	c	c
	24	c	c	c	c
	25*	20 min	2.5 min	saline	2.5 min
	26*	c	c	c	c
	27	c	c	c	c
	28*	saline	saline	saline	saline
	29*	0.75 d-A	0.5 d-A	0.75 d-A	0.75 d-A
	30	c	c	c	c
	31	c	c	c	c
	32	c	c	c	c
	33*	48 hr dep	48 hr dep	48 hr dep	48 hr dep
	34*	72 hr dep	72 hr dep	72 hr dep	72 hr dep
		+0.25 d-A	+0.25 d-A	+1.5 d-A	+1.5 d-A
	35	c	c	c	c
	36	c	c	c	c
	37	c	c	c	c
IB	38*	saline	saline	saline	saline
	39*	1.5 d-A	0.25 d-A	1.5 d-A	1.5 d-A
	40	c	c	c	c
	41	c	c	c	c
	42*	c	c	c	c
	43*	2.5 min	2.5 min	2.5 min	2.5 min
		+0.75 d-A	+0.75 d-A	+0.75 d-A	+0.75 d-A
	44	c	c	c	c
	45*	c	c	c	c
	46*	48 hr dep	48 hr dep	48 hr dep	48 hr dep
	47*	72 hr dep	72 hr dep	72 hr dep	72 hr dep
		+1.5 d-A	+1.5 d-A	+0.75 d-A	+0.75 d-A
	48	c	c	c	c
	49	c	c	c	c
	50*	c	c	c	c
	51*	2.5 min	2.5 min	2.5 min	2.5 min
		+0.25 d-A	+0.25 d-A	+1.5 d-A	+1.5 d-A
	52	c	c	c	c
	53	c	c	c	c
	54*	48 hr dep	48 hr dep	48 hr dep	48 hr dep
	55*	72 hr +0.75	72 hr +0.75	74 hr +0.25	74 hr +0.25
		d-A	d-A	d-A	d-A

(Continues)

Day	1 & 2	3 & 4	5 & 6	7 & 8
56	c	c	c	c
57	c	c	c	c
58	c	c	c	c
59*	saline	saline	saline	saline
60*	2.5 min +1.5 d-A	2.5 min +1.5 d-A	2.5 min +0.25 d-A	2.5 min +0.25 d-A
61	c	c	c	c
62	c	c	c	c
63	c	c	c	c
64*	48 hr dep	48 hr dep	48 hr dep	48 hr dep
65*	72 hr dep +0.5 d-A	72 hr dep +0.5 d-A	74 hr dep +0.5 d-A	72 hr dep +0.5 d-A
66	c	c	c	c
67	c	c	c	c
68	c	c	c	c
69*	c	c	c	c
70*	2.5 min +0.5 d-A	2.5 min +0.5 d-A	2.5 min +0.5 d-A	2.5 min +0.5 d-A
71	c	c	c	c
72*	c	c	c	c
73*	48 hr dep	48 hr dep	48 hr dep	48 hr dep
74*	72 hr dep	72 hr dep	72 hr dep	72 hr dep
75	c	c	c	c

REFERENCES

1. S. A. Corson and E. O'Leary Corson, Pavlovian conditioning as a method for studying the mechanisms of action of minor tranquilizers. Neuropsychopharmacology, Proceedings of the Intern. Congr. of College Instruction and Neuropsychopharmacology, 5th, Washington, D. C., pp. 857-78, 1967.

2. B. F. Skinner, The Behavior of Organisms, Appleton-Century-Crofts, New York, 1938.

3. C. B. Ferster and B. F. Skinner, Schedules of Reinforcement, Appleton-Century-Crofts, New York, 1957.

4. F. S. Keller and W. N. Schoenfeld, Principles of Psychology, Appleton-Century-Crofts, New York, 1950.

5. J. G. Holland and B. F. Skinner, The Analysis of Behavior: A Program for Self-Instruction, McGraw-Hill, New York, 1961.

6. E. P. Reese, Experiments in Operant Behavior, Appleton-Century-Crofts, New York, 1964.

7. C. A. Catania, ed., Contemporary Research in Operant Behavior, Scott, Foresman and Co., Glenview, Illinois, 1968.

8. W. K. Honig, ed., Operant Behavior: Areas of Research and Application, Appleton-Century-Crofts, New York, 1966.

9. I. Goldiamond (personal communication), to appear in I. Goldiamond, and D. M. Thompson.

10. I. Goldiamond (personal communication), to appear in I. Goldiamond and D. M. Thompson. For earlier version, see "The Operant Paradigm," in I. Goldiamond and J. Dyrud, Some applications and implications of behavioral analysis for psychotherapy, Res. Psychotherap., 3, 67–69 (1968).

11. V. G. Laties, B. Weiss, R. L. Clark, and M. D. Reynolds, Overt "mediating" behavior during temporally spaced responding, J. Exptl. Anal. Behav., 8, 107–116 (1965).

12. M. P. Wilson and F. S. Keller, On the selective reinforcement of spaced responses, J. Comp. Physiol. Psychol., 46, 190–193 (1953).

13. A. Herz, Drugs and the conditioned avoidance response, Intern. Rev. Neurobiol., 2, 229–277 (1960).

14. M. Sidman, Avoidance conditioning with brief shock and no exteroceptive warning signal, Science, 118, 157–158 (1953).

15. B. A. Campbell and D. Kraeling, Response strength as a function of drive level and amount of drive reduction, J. Exptl. Phychol., 45, 97–101 (1953).

16. G. A. Kimble, Shock intensity and avoidance learning, J. Comp. Physiol. Psychol., 48, 281–284 (1955).

17. V. H. Denenberg, Interactive effects of infantile and adult shock levels upon learning, Psychol. Reports, 5, 357–364 (1959).

18. J. J. Boren, M. Sidman, and R. J. Herrnstein, Avoidance, escape, and extinction as functions of shock intensity, J. Comp. Physiol. Psychol., 52, 420–425 (1959).

19. F. W. Huff, T. P. Piantanida, and G. L. Morris, Free operant avoidance responding as a function of serially presented variations of UCS intensity, Psychonomic Science, 8, 111–112 (1967).

20. K. E. Moyer and J. H. Korn, Effect of UCS intensity on the acquisition and extinction of an avoidance response, J. Exptl. Psychol., 67, 352–359 (1964).

21. S. Levine, UCS intensity and avoidance learning, J. Exptl. Psychol., 71, 163-164 (1966).

22. R. W. Powell, The effect of shock intensity upon responding under a multiple-avoidance schedule, J. Exptl. Anal. Behav., 14, 321-329 (1970).

23. M. R. D'Amato and D. Schiff, Long-term discriminated avoidance performance in the rat, J. Comp. Physiol. Psychol., 57, 123-126 (1964).

24. M. R. D'Amato, D. Keller, and G. Biederman, Discriminated avoidance learning as a function of parameters of discontinuous shock, J. Exptl. Psychol., 70, 543-548 (1965).

25. M. R. D'Amato, D. M. Keller, and L. DiCara, Facilitation of discriminated avoidance learning by discontinuous shock, J. Comp. Physiol. Psychol., 58, 344-349 (1964).

26. B. Weiss and V. G. Laties, Drug effects on the temporal patterning of behavior, Federation Proc., 23, 801-807 (1964).

27. H. Barry, III, W. J. Kinnard, Jr., N. Watzman, and J. P. Buckley, A computer-oriented system for high-speed recording of operant behavior, J. Exptl. Anal. Behav., 9, 163-171 (1966).

28. R. M. Herrick and J. S. Denelsbeck, A system for programming experiments and for recording and analyzing data automatically, J. Exptl. Anal. Behav., 6, 631-635 (1963).

29. L. S. Seiden, R. Schoenfeld, and D. Domize, A system for the recording and analysis of inter-response time data using an AM tape recorder and digital computers, J. Exptl. Anal. Behav., 12, 289-292 (1969).

30. P. B. Dews, Analysis of effects of psychopharmacological agents in behavioral terms, Federation Proc., 17, 1024-1030 (1958).

31. P. B. Dews, Studies on Behavior, I. Differential sensitivity to pentobarbital of pecking performance in pigeons depending on the schedule of reward, J. Pharmacol. Exptl. Therap., 113, 393-401 (1955).

32. P. B. Dews, Studies on Behavior, IV. Stimulant actions of methamphetamine. J. Pharmacol. Exptl. Therap., 122, 137-147 (1958).

33. R. I. Schoenfeld and L. S. Seiden, Effect of α-methyltyrosine on operant behavior and brain catecholamine levels, J. Pharmacol. Exptl. Therap., 167, 319-327 (1969).

<u>34.</u> M. G. Waller and P. F. Waller, Effects of chlorpromazine on appetitive and aversive components on a multiple schedule, J. Exptl. Anal. Behav., 5, 259-264 (1962).

<u>35.</u> L. Cook and A. C. Catania, Effects of drugs on avoidance and escape behavior, Federation Proc., 23, 818-835 (1964).

<u>36.</u> L. Cook and R. T. Kelleher, Drug effects on the behavior of animals, Ann. New York Acad. Sci., 96, 315-335 (1962).

<u>37.</u> L. Cook and R. T. Kelleher, Effects of drugs on behavior, Ann. Rev. Pharmacol., 3, 205-222 (1963).

<u>38.</u> R. T. Kelleher and W. H. Morse, Escape behavior and punished behavior, Federation Proc., 23, 808-817 (1964).

<u>39.</u> P. B. Dews and W. H. Morse, Behavioral pharmacology, Ann. Rev. Pharmacol., 1, 145-174 (1961).

<u>40.</u> L. Cook and E. F. Weidley, Behavioral effects of some psychopharmacological agents, Ann. New York Acad. Sci., 66, 740-752 (1957).

<u>41.</u> S. Courvoisier, J. Fournel, R. Ducrot, M. Kolsky, and P. Koetschet, Propriétés Pharmacodynamiques du Chlorhydrate de Chloro-3 (Diméthylamine-3'-propyl)-10-Phénothiazine (4.560 R. P.) Arch. Intern. Pharmacodyn., 92, 305-361 (19531.

<u>42.</u> R. E. Miller, J. V. Murphy, and I. A. Mirsky, The effect of chlorpromazine on fear-motivated behavior in rats, J. Pharmacol. Exptl. Therap., 120, 379-387 (1957).

<u>43.</u> H. F. Hunt, Methods for studying the behavioral effects of drugs, Ann. Rev. Pharmacol., 1, 125-144 (1961).

<u>44.</u> L. A. Low, M. Eliasson, and C. Kornetsky, Effect of chlorpromazine on avoidance acquisition as a function of CS-US interval length, Psychopharmacologia, 10, 148-154 (1966).

<u>45.</u> O. S. Ray and L. W. Bivens, Performance as a function of drug, dose and level of training, Psychopharmacologia, 10, 103-109 (1966).

<u>46.</u> O. S. Ray and L. W. Bivens, Chlorpromazine and amphetamine effects on three operant and on four discrete trial reinforcement schedules, Psychopharmacologia, 10, 32-43 (1966).

<u>47.</u> B. A. Doty and L. A. Doty, Facilitative effects of amphetamine on avoidance conditioning in relation to age and problem difficulty, Psychopharmacologia (Berlin), 9, 234-241 (1966).

48. A. K. Myers, Effects of CS intensity and quality in avoidance conditioning, J. Comp. Physiol. Psychol., 55, 57–61 (1962).

49. D. R. Meyer, C. Chungsoo, and N. S. Wesemann, On problems of conditioning discriminated lever-press avoidance responses, Psychological Rev., 67, 224–228 (1960).

50. F. R. Brush, The effects of intertrial interval on avoidance learning in the rat, J. Comp. Physiol. Psychol., 55, 888–892 (1962).

51. J. Pearl, Intertrial interval and acquisition of a lever press avoidance response, J. Comp. Physiol. Psychol., 56, 710–712 (1963).

52. J. V. Murphy and R. E. Miller, Spaced and massed practice with a methodological consideration of avoidance conditioning, J. Exptl. Psychol., 52, 77–81 (1956).

53. C. Y. Nakamura and N. H. Anderson, Avoidance behavior differences within and between strains of rats, J. Comp. Physiol. Psychol., 55, 740–747 (1962).

54. S. S. Tenen, The effects of p-chlorophenylalanine, a serotonin depletor, on avoidance acquisition, pain sensitivity and related behavior in the rat, Psychopharmacologia, 10, 204–219 (1967).

55. J. A. Harvey and C. E. Lints, Lesions in the medial forebrain bundle: Delayed effects on sensitivity to electric shock, Science, 148, 250–252 (1965).

56. W. O. Evans, A new technique for the investigation of some analgesic drugs on a reflexive behavior in the rat, Psychopharmacologia, 2, 318–325 (1961).

57. B. Weiss and V. G. Laties, Fractional escape and avoidance on a titration schedule, Science, 128, 1575–1576 (1958).

58. B. Weiss and V. G. Laties, Titration behavior on various fractional escape programs, J. Exptl. Anal. Behav., 2, 227–248 (1959).

59. B. Weiss and V. G. Laties, Changes in pain tolerance and other behavior produced by salicylates, J. Pharmacol. Exptl. Therap., 131, 120–129 (1961).

60. H. F. Hunt and J. V. Brady, Some effects of electro-conculsive shock on a conditioned emotional response: The effect of post-ECS extinction on the reappearance of the response, J. Comp. Physiol. Psychol., 45, 589–596 (1951).

61. Z. Annau and L. J. Kamin, The conditioned emotional response as a function of intensity of the US, J. Comp. Physiol. 54, 428–432 (1961).

62. S. S. Tenen, Recovery time as a measure of CER strength. Effects of benzodiazepines, amobarbital, chlorpromazine, and amphetamine, Psychopharmacologia, 12, 1-17 (1967).

63. W. K. Estes and B. F. Skinner, Some quantitative properties of anxiety, J. Exptl. Psychol., 29, 390-400 (1941).

64. J. V. Brady, A comparative approach to the evaluation of drug effects upon affective behavior, Ann. N. Y. Acad. Sci., 64, 632-643 (1956).

65. J. B. Appel, Drugs, shock intensity and the CER, Psychopharmacologia, 4, 148-153 (1963).

66. R. S. Yamahiro, E. C. Bell, and H. E. Hill, The effects of reserpine on a strongly conditioned emotional response, Psychopharmacologia, 2, 197-202 (1961).

67. H. E. Hill, E. C. Bell, and A. Wikler, Reduction of conditioned suppression: Actions of morphine compared with those of amphetamine, pentobarbital, nalorphine, cocaine, LSD-25 and chlorpromazine, Arch. Intern. Pharmacodyn., 165, 212-226 (1967).

68. I. Geller, The acquisition and extinction of conditioned suppression as a function of the baseline reinforcer, J. Exptl. Anal. Behav., 3, 235-240 (1960).

69. L. Stein, M. Sidman, and J. V. Brady, Some effects of two temporal variables on conditioned suppression, J. Exptl. Anal. Behav., 1, 153-162 (1958).

70. J. A. Harvey, C. E. Lints, L. E. Jacobson, and H. F. Hunt, Effects of lesions in the septal area on conditioned fear and discriminated instrumental punishment in the albino rat, J. Comp. Physiol. Psychol., 59, 37-48 (1965).

71. D. Blackman, Effects of drugs on conditioned "anxiety," Nature, 217, 769-770 (1968).

72. A. Weissman, Differential drug effects upon a three-ply multiple schedule of reinforcement, J. Exptl. Anal. Behav., 2, 271-287 (1959).

73. O. S. Ray, The effects of tranquilizers on positively and negatively motivated behavior in rats, Psychopharmacologia, 4, 326-342 (1963).

74. T. Thompson and C. R. Schuster, Behavioral Pharmacology, Prentice-Hall, Englewood Cliffs, N. J., 1968.

75. B. Weiss and A. Heller, Methodological problems in evaluating the role of cholinergic mechanisms in behavior, Federation Proc., 28, 135-146 (1967).

76. B. Weiss and V. Laties, Behavioral pharmacology and toxicology, Ann. Rev. Pharmacol., 9, 297–326 (1969).

77. L. S. Seiden, R. M. Brown, and A. J. Lewy, Brain catecholamines and conditioned behavior: Mutual interactions, James E. P. Toman Memorial Symposium. Hector Sabelli, ed.

78. N. W. Heimstra and A. McDonald, Social influence on the response to drugs. III. Response to amphetamine sulfate as a function of age, Psychopharmacologia, 3, 212–218 (1962).

79. M. R. D'Amato and H. Jagoda, Age, sex and rearing conditions as variables in simple brightness discrimination, J. Comp. Physiol. Psychol., 53, 261–263 (1960).

80. J. R. Sloanaker, The normal activity of the white rat at different ages, J. Comp. Neurol. Psychol., 17, 342–359 (1907).

81. J. L. Fuller, Variation in effects of chlorpromazine in three strains of mice, Psychopharmacologia, 8, 408–414 (1966).

82. B. D. Gupta and K. Gregory, The effect of drugs and their combinations on the rearing response in two strains of rats, Psychopharmacologia, 11, 365–371 (1967).

83. G. C. Walters, J. Pearl, and J. V. Rogers, The gerbil as a subject in behavioral research, Psychol. Rep., 12, 315–318 (1963).

84. D. K. Candland, B. Faulds, D. B. Thomas, and M. H. Candland, The reinforcing value of gentling, J. Comp. Physiol. Psychol., 53, 55–58 (1960).

85. H. C. Agrawal, M. W. Fox, and W. A. Himwich, Neurochemical and behavioral effects of isolation-rearing in the dog, Life Sci., 6, 71–78 (1967).

86. W. F. Angermeir, Some basic aspects of social reinforcement in albino rats, J. Comp. Physiol. Psychol., 53, 364–367 (1960).

87. M. R. Denny and R. G. Wiesman, Avoidance behavior as a function of length of nonshock confinement, J. Comp. Physiol. Psychol., 58, 252–257 (1964).

88. E. N. Greenblatt and A. C. Osterberg, Correlations of activating and lethal effects of excitatory drugs in grouped and isolated mice, J. Pharmacol. Exptl. Therap., 131, 115–119 (1961).

89. K. R. Hughes and J. P. Zubek, Effect of glutamic acid on the learning ability of bright and dull rats: II. Duration of the effect, Can. J. Psychol., 11, 182–184 (1957).

90. K. R. Hughes and J. P. Zubek, Effect of glutamic acid on the learning ability of bright and dull rats: I. Administration during infancy, Can. J. Psychol., 10, 132–138 (1956).

91. W. J. Meyers, Effects of different intensities of postweaning shock and handling on the albino rat. J. Genet. Psychol., Child Behav. Animal Behav. Comp. Psychol., 106, 51–58 (1965).

92. H. Barry, III, Habituation to handling as a factor in retention of maze performance in rats, J. Comp. Physiol. Psychol., 50, 366–367 (1957).

93. B. Gertz, The effects of handling at various age levels on emotional behavior of adult rats, J. Comp. Physiol. Psychol., 50, 613–616 (1957).

94. H. H. Reiter, Effects of noise on discrimination reaction time, Perceptual and Motor Skills, 17, 418 (1963).

95. H. H. Reynolds, Effects of rearing and habitation in social isolation on performance of an escape task, J. Comp. Physiol. Psychol., 56, 520–525 (1963).

96. J. Rosen, Effects of early social experience upon behavior and growth in the rat, Child Development (3), 35, 993–998 (1964).

97. J. H. Pirch and R. H. Rech, Effect of isolation on α-methyltyrosine-induced behavioral depression, Life Sci., 7, 173–182 (1968).

98. J. H. Stock, Some effects at maturity of gentling, ignoring, or shocking rats during infancy, J. Abnormal Psychol. Social Psychol., 51, 412–414 (1955).

99. D. D. Thiessen, J. F. Zolman, and D. A. Rodgers, Relation between adrenal weight, brain cholinesterase activity, and hole-in-wall behavior of mice under different living conditions, J. Comp. Physiol., 55, 186–190 (1962).

100. J. A. Dinsmoor, Variable-interval escape from stimuli accompanied by shocks, J. Exptl. Anal. Behav., 5, 41–47 (1962).

101. C. B. Ferster, J. B. Appel, and R. A. Hiss, The effect of drugs on a fixed-ratio performance suppressed by a pre-timeout stimulus, J. Exptl. Anal. Behav., 5, 73–88 (1962).

102. V. G. Laties and B. Weiss, Influence of drugs on behavior controlled by internal and external stimuli, J. Pharmacol. Exptl. Therap., 152, 388–396 (1966).

103. C. L. Roberts, M. H. Marks, and G. Collier, Light onset and light offset as reinforcers for the albino rats, J. Comp. Physiol. Psychol., 51, 575-579 (1958).

104. R. E. Ulrich, W. C. Holz, and N. H. Azrin, Stimulus control of avoidance behavior, J. Exptl. Anal. Behav., 7, 129-133 (1964).

105. L. Stein, M. Sidman, and J. V. Brady, Some effects of two temporal variables on conditioned suppression, J. Exptl. Anal. Behav., 1, 153-162 (1958).

106. W. C. Stebbins and J. M. Miller, Reaction time as a function of stimulus intensity for the monkey, J. Exptl. Anal. Behav., 7, 309-312 (1964).

107. A. K. Myers, Onset vs. termination of stimulus energy as the CS in avoidance conditioning and pseudoconditioning, J. Comp. Physiol. Psychol., 53, 72-78 (1960).

108. C. A. Berry and L. G. Stark, Modification of conditional behavior by prior experience - Effects of scopolamine, Psychopharmacologia, 7, 409-415 (1965).

109. J. L. Andreass and P. M. Whalen, Some physiological correlates of learning and overlearning, Psychophysiology, 3, 406-413 (1967).

110. W. W. Cumming and W. N. Schoenfeld, Some data on behavior reversibility in a steady state experiment, J. Exptl. Anal. Behav., 2, 87-90 (1959).

111. J. L. Fozard, Trial spacing in instrumental running, Psychol. Rep., 18, 623-630 (1966).

112. H. M. B. Hurwitz, Method for discriminative avoidance training, Science, 145, 1070-1071 (1964).

113. S. Levine and S. J. England, Temporal factors in avoidance learning, J. Comp. Physiol. Psychol., 53, 282-283 (1960).

114. A. K. Myers, Avoidance learning as a function of several training conditions and strain differences in rats, J. Comp. Physiol. Psychol., 52, 381-386 (1959).

115. J. B. Appel and H. M. B. Hurwitz, Studies in light-reinforced behavior. IV: Effects of apparatus familiarization, Psychol. Rep., 5, 355-356 (1959).

116. H. S. Hoffmann and M. Fleshler, Startle reaction: Modification by background acoustic stimulation, Science, 141, 928-930 (1963).

117. R. C. Beck and J. F. McLean, Effect of schedule of reinforcement and stomach loads on bar pressing by thirsty rats, J. Comp. Physiol. Psychol., 63, 530-533 (1967).

118. D. Belanger and S. M. Felman, Effects of water-deprivation upon heart rate and instrumental activity in the rat, J. Comp. Physiol. Psychol., 55, 220-225 (1962).

119. F. W. Finger, L. S. Reid, and M. H. Weasner, Activity changes as a function of reinforcement under low drive, J. Comp. Physiol. Psychol., 53, 385-387 (1960).

120. R. C. Miles, Effect of food deprivation on manipulatory reactions in cat, J. Comp. Physiol. Psychol., 55, 358-362 (1962).

121. H. Brown and R. K. Richards, An interaction between drug effects and food reinforced "social" behavior in pigeons, Arch Intern. Pharmacodyn., 164, 286-293 (1966).

122. N. H. Azrin, W. C. Holz, and D. Hake, Intermittent reinforcement by removal of a conditioned aversive stimulus, Science, 136, 781-782 (1962).

123. F. C. Clark, Some observations on the adventitious reinforcement of drinking under food reinforcement, J. Exptl. Anal. Behav., 5, 61-63 (1962).

124. P. B. Dews, Studies on responding under fixed-interval schedules of reinforcement: The effects on the pattern of responding of changes in requirements at reinforcement, J. Exptl. Anal. Behav., 12, 191-201 (1969).

125. J. Faidherbe, M. Richelle, and J. Schlag, Nonconsumption of the reinforcer under drug action, J. Exptl. Anal. Behav., 5, 521-524 (1962).

126. N. Guttman, Equal-reinforcement values for sucrose and glucose solutions compared with equal-sweetness values, J. Comp. Physiol. Psychol., 47, 358-361 (1954).

127. N. Guttman, Operant conditioning, extinction and periodic reinforcement in relation to concentrations of sucrose used as reinforcing agent, J. Exptl. Psychol., 46, 213-224 (1953).

128. M. F. Halasz, Stability of conditioned delay behavior computed by transient response to perturbation of reward contingency, Am. Zoologist, 6, 543-544 (1966).

129. M. F. Halasz, A behavioral evoked response: Probing the stability of delayed conditioned approach (albino rat) with impulse-like changes of reinforcement schedule, Am. Zoologist, 7, 787-788 (1967).

130. H. M. B. Hurwitz, Conditioned responses in rats reinforced by light, Brit. J. Animal Behav., 4, 31-33 (1956).

131. H. M. B. Hurwitz and J. B. Appel, Light-onset reinforcement as a function of the light-dark maintenance schedule for the hooded rat, J. Comp. Physiol. Psychol., 52, 710-712 (1959).

132. R. T. Kelleher and L. R. Gollub, A review of positive conditioned reinforcements, J. Exptl. Anal. Behav., 5, 543-597 (1962).

133. M. J. Homzie and L. E. Ross, Runway performance following a reduction in the concentration of a liquid reward, J. Comp. Physiol. Psychol., 55, 1029-1033 (1962).

134. W. B. Pavlik and W. E. Reynolds, Effect of deprivation schedule and reward magnitude on acquisition and extinction performance, J. Comp. Physiol. Psychol., 56, 452-455 (1963).

135. A. C. Pereboom and B. M. Crawford, Instrumental and competing behavior as a function of trials and reward magnitude, J. Exptl. Psychol., 56, 82-85 (1958).

136. W. C. Stebbins, Response latency as a function of amount of reinforcement, J. Exptl. Anal. Behav., 5, 305-308 (1962).

137. W. C. Stebbins, P. B. Mead, and J. M. Martin, The relation of amount of reinforcement to performance under a fixed-interval schedule, J. Exptl. Anal. Behav., 2, 351-355 (1959).

138. H. M. Hanson, E. H. Campbell, and J. J. Witoslawski, FI length and performance on an FI/FR chain schedule of reinforcement, J. Exptl. Anal. Behav., 5, 331-334 (1962).

139. B. F. Skinner, Diagramming schedules of reinforcement, J. Exptl. Anal. Behav., 1, 67-68 (1958).

140. W. C. Stebbins and R. N. Lanson, Response latency as a function of reinforcement schedule, J. Exptl. Anal. Behav., 5, 229-304 (1962).

141. H. O. Doerr, A method to establish electrically good contacts with unrestrained rats, Psychophysiology, 3, 316-317 (1967).

142. M. R. D'Amato and J. Fazzaro, Discriminated lever-press avoidance learning as a function of type and intensity of shock, J. Comp. Physiol. Psychol., 61, 313-315 (1966).

Chapter 3

MICROTUBULES AND MICROTUBULAR PROTEIN

Frederick E. Samson, Jr., Dianna Ammons Redburn, and Richard H. Himes
Department of Physiology and Cell Biology
and
Department of Biochemistry
University of Kansas
Lawrence, Kansas

I. INTRODUCTION

Microtubules are among the most numerous "organelles" in the ultra-structure of neurons and are universally present in eukaryotes (1, 2). They occur in the mitotic spindle of dividing cells, cilia, flagella, sperm tails, and are abundant in neuronal axons and dendrites. Representative

electron micrographs showing the abundance and morphology of microtubules in crayfish ventral nerve cord are presented in Figs. 1-3. In a
few cases, direct counts of the number of microtubules have been made
and the results indicate that they are more numerous in nonmyelinated
than myelinated axons (3, 4). They are also more abundant in the axons

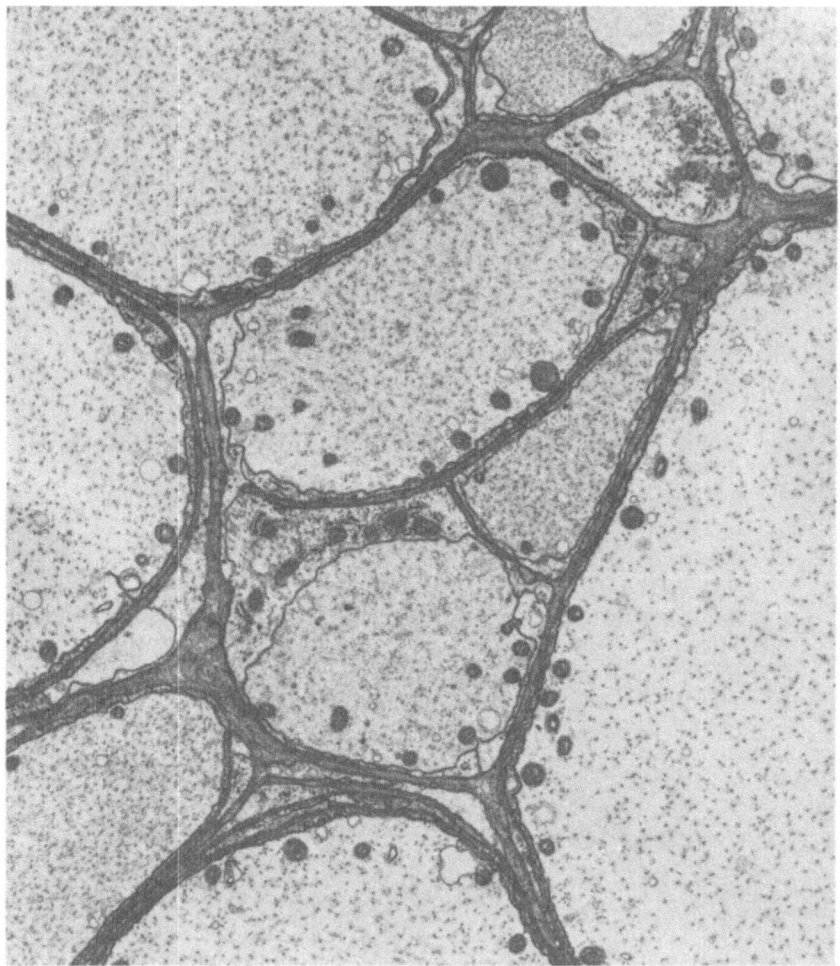

FIG. 1. Electron micrograph illustrating the general ultrastructure
of the crayfish ventral nerve cord in cross section. Microtubules are
generally oriented parallel to the long axis of the axon and appear as small
dots scattered throughout the axoplasm. Mitochondria are located near the
periphery of the axons. x 10,000

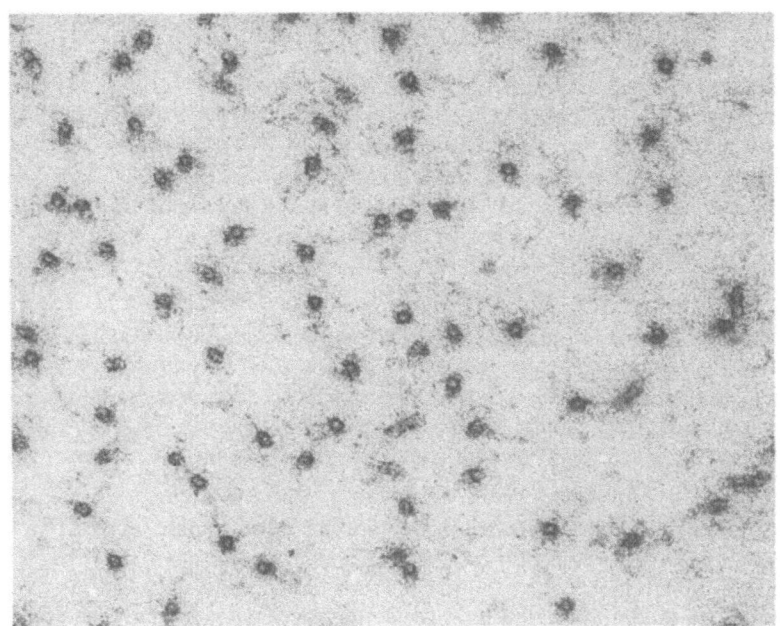

FIG. 2. Enlargement of cross section showing electron lucent center of microtubules and wispy filaments radiating from their exterior surface. x 80,000

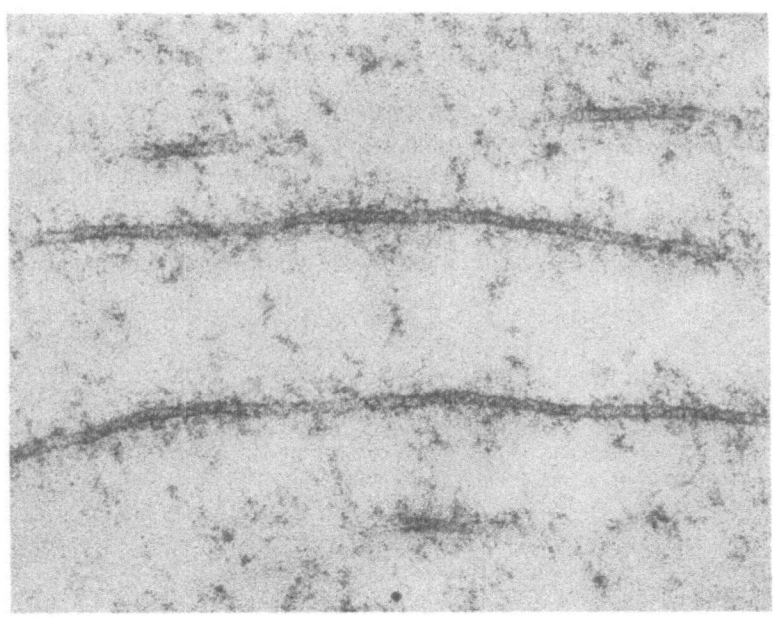

FIG. 3. Microtubules viewed in longitudinal section of crayfish ventral nerve cord. x 80,000

of immature animals than in mature animals (5). An extensive survey of microtubules in rabbit vagus nerve nonmyelinated axons and in crayfish ventral cord axons revealed a considerable numerical variation among axons, even from the same animal. The average number per square micron is 40 in rabbit vagus and 26 in crayfish ventral cord (3).

Microtubules are characterized by their tubular appearance and possess remarkably uniform properties over a diversity of cell types. They have an outside diameter of approximately 250 Å, inside diameter of 130 Å, and a wall thickness of roughly 60 Å (2). The tube length is usually difficult to ascertain but in some cases they are clearly several microns long (2).

Microtubules are composed of protein subunits which chemically are similar across the great diversity of cell types where they exist (2). This protein has been isolated by the procedures described below and the name "tubulin" (6) is receiving increasing use and is adopted here. The subunits are polymerized by forces that are not understood, but the susceptibility of microtubules to high pressure (7, 8) and cooling (9) and the stabilization by D_2O (10, 11) suggests that weak bonding is involved. Electron microscopy of transverse sections indicate that 10-14 subunits form the perimeter of the tubule (12).

Tubulin is characterized principally by its ability to bind colchicine (Fig. 4), indeed it is often referred to as colchicine-binding protein. It binds GTP and Mg and has an average molecular weight of 110-120,000 D (13-15). When isolated by the usual procedures the subunits are primarily in the dimer form (16). The monomer is believed to be somewhat oblong in shape (40 x 60 Å) (17). The molecule is acidic and the aminoacid composition is similar for colchicine-binding protein isolated from widely different sources. There does not appear to be any lipid associated with the subunit but a small amount of carbohydrates (1%) has been reported (18).[1]

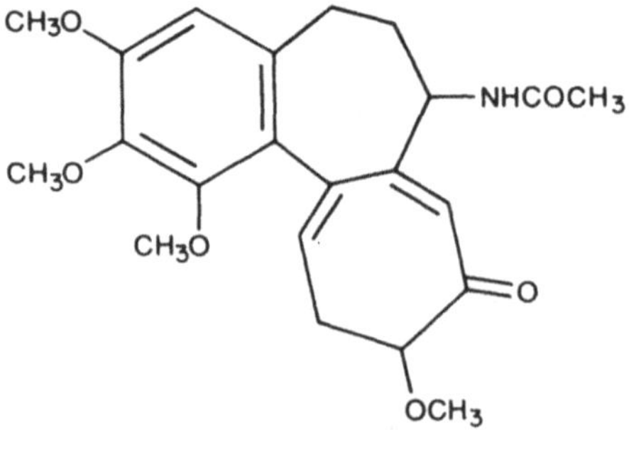

FIG. 4. Chemical formula of colchicine.

[1]All numbered footnotes refer to Notes Added in Proof, p. 146.

Refined techniques have recently revealed several types of monomers which differ slightly in molecular weight and electrophoretic properties (14-15). It is not clear at the present time whether a given microtubule is composed of only one type of subunit or varying proportions of more than one type. The degree of dissimilarity of tubulin isolated from cells representing the phylogenetic scale has not been extensively catalogued. However, it does appear that tubulin was a highly conservative protein in evolution.

Microtubules occur as part of larger systems. Indeed even cytoplasmic tubules running lengthwise in the axoplasm which appear as single entities with traditional electron microscopic procedures, may be coordinated within a larger system. Axonal microtubules are ensheated with a material which stains with ruthenium red (19) and lathanum hydroxide (20), indicating that it is composed of an anionic polyelectrolyte and possibly acid mucopolysaccharide. The ensheathing material appears to be part of a larger, grid-like system resembling a spider web with the microtubules located at the bonding points (21). A number of reports have appeared which claim that distinct cross-bridges exist between microtubules and other cellular structures. The most striking example of this is seen in studies by Smith on the lamprey in which clusters of synaptic vesicles are attached to single microtubules by distinct cross-bridges (22).

In many nonneuronal cells, microtubules are also linked by cross-bridges into patterns such as the nine doublets of cilia axonemes (23), the interlocking spirals in heliozoan axopods (24), and the circular groupings in the nematocyst of Suctoria (25). In other cases linking cross-bridges are less prominent and possibly are to unstable to be revealed by the current electron microscopy procedures.

Microtubules are generally unstable and disappear with a variety of treatments. Behnke and Forer noted differences in stability of microtubules to cooling, pepsin digestion, colchicine, and warming (26). From these observations they concluded that at least four classes of microtubules exist (26). It is not known whether the different stabilities result from variations in the composite protein or to the environmental conditions surrounding the microtubules, such as the nature of the ensheathing material. The variation in stability is also seen in microtubules from different axons. For example, colchicine depolymerizes microtubules in the rabbit vagus fibers (3) but has no readily apparent effect on the morphology of the microtubules in crayfish ventral cord axons (27). The apparent stability of the microtubules may depend upon the rate of exchange of polymerized and depolymerized subunits. If the colchicine interacts with the depolymerized but not the polymerized subunits, it will more effectively disrupt the tubules in those cases where the exchanges are rapid.

A sizable list of functions has been suggested for microtubules or microtubular systems. Suggestions range from the very speculative to the reasonably secure. One provocative idea which takes into account that the microtubules are continuous over a long stretch of the axons or dendrites, proposes that microtubules act as a communication system conducting "signals" from one place to another within the cell. Microtubular arrays are common in sensory receptors and they may be intimately involved in sensory transduction (28). One function, cytoskeletal support, has been experimentally demonstrated. Indeed, it has been shown that the integrity of the microtubules is necessary for the individual shape of many cells and the rigidity of elongated cellular processes (29, 30). An additional suggestion is that microtubules or microtubular systems are concerned with the coupling of metabolically produced chemical energy to mechanical events, in other words, they are "chemo-mechanical energy couplers." The persuasion for this view is their occurrence in motile systems such as flagella, cilia, and sperm tails.

In the nervous system the movement of materials from the cell body out and along the axons, axoplasmic transport, may be a special case of coupling metabolic energy to a mechanical process. Work from many laboratories has shown that the integrity of the axonal microtubules is critical in axoplasmic transport (31, 32). The transport of pigment granules in melanocytes is sufficiently similar to the axonal transport of neuronal constituents that some lessons can be taken from the detailed studies on the melanocyte. The involvement of microtubules in movement of the pigment granules is inferred from (a) the physical relationship of the granules to microtubules during their transport, and (b) the inhibition of transport by agents which react with tubulin and disintegrate microtubules, such as vinblastine and colchicine (33). The conclusions from these observations are that the microtubules define the channels and possibly apply the motive force for the granule movements (33, 34). Finally, there is evidence that tubulin is a component in neuronal membranes (35, 36) and may be involved in the release of secretory products and neurotransmitters (37, 38). We would like to emphasize that the various theories concerning the function of microtubules are not mutually exclusive.

II. ASSAY PROCEDURE

The standard assay for tubulin is based on its unique ability to bind colchicine. The structure of colchicine is shown in Fig. 4. Colchicine is an alkaloid found in the autumn crocus, Colchicum autumnae, which has been used therapeutically since the sixth century A.D. for the relief of pain from gouty arthritis (39). In more recent times, it has been used extensively as a potent mitotic inhibitor. Experiments by Taylor (40, 41) and others indicated that colchicine binds to a specific protein found both in dividing and nondividing cells. By using colchicine as a marker, Borisy and Taylor were able to isolate, purify, and characterize what appeared to be a single species of colchicine-binding protein (41). The direct correlation between the amount of colchicine-binding activity in a tissue or cell type with the occurrence of microtubules provided indirect evidence that colchicine-binding protein is a major constituent of microtubules (41). Further it was found that the time-dependent appearance of a 6S colchicine-binding protein in solution was associated with the disappearance of the central pair of microtubules in suspensions of isolated sea urchin sperm tails (42). Some physiological effects of colchicine can also be explained on the basis of its action on microtubules. For example: colchicine stops mitosis by disrupting mitotic spindle microtubules (43). The action of colchicine on the symptoms of gout is not as well understood; however, Malawista (44) suggests that colchicine stops phagocytosis of urate crystals in gouty joints by disrupting macrophage microtubules.

The effect of colchicine in vivo varies among different systems. Many microtubules (e.g., those in rabbit vagus fiber) depolymerize in the presence of colchicine (3); others (e.g., those in crayfish ventral cord axons) do not (27). Colchicine treatment also affects other cellular substructures; some examples are: (a) the fragmentation of the Golgi apparatus and proliferation of 60–80 Å cytoplasmic fibrils in interphase HeLa cells (45); (b) an increase in the electron density of nuclear membranes and the number of nuclear pores, and abnormally prominent clusters of 100 Å fibrils in neurons (46). Further, colchicine has also been shown to inhibit RNA synthesis (47). Although these changes may be secondary to colchicine interaction with tubulin, they illustrate that the impairment of a function by colchicine may not necessarily imply the involvement of microtubules.

It should be emphasized that the evidence for the specificity of the binding of colchicine to microtubule protein is based on in vitro studies on supernatant fractions. Microtubular protein in this fraction presumably represents the 120,000 mol. wt. dimeric subunits from microtubules which have been depolymerized during homogenization and centrifugation in addition to subunits from any preexisting pool of subunits. Colchicine also binds to particulate fractions (48); however, the nature of the particulate colchicine binding protein is unknown. Not all of the colchicine binding can be accounted for by subunits entrapped in particulate structures, since extensive washing of the particulate fraction does not remove all of the colchicine binding activity (48). This suggests that tubulin may form an integral part of membrane structure. However, tubulin in its native state readily aggregates even under optimal conditions and therefore must be regarded as a potential contaminant of all particulate fractions. At this point, the exact nature of the particulate colchicine binding protein is unresolved. Whether this protein is a functionally important part of cellular membranes or simply an aggregation artifact produced during preparation procedures, remains to be determined.

Little is known about the mechanism of the colchicine binding. It is generally accepted that there is one binding site per 120, 000 mol. wt. of the protein and that the binding site is active only when the protein is in its native state (16). The colchicine is noncovalently bound and its apparent structure remains unchanged after the relatively long and obviously complex binding mechanism is complete (49). Borisy and Taylor (41) reported binding constants of 2.3 X 10^6 liter moles^{-1} in vitro and 1.2 X 10^6 liter moles^{-1} in vivo. More exact characterization of the binding mechanism has been hampered by the unstable nature of the native protein. The half-life of tubulin from porcine brain is about 11 hr (16) and from chick embryo brain, the half-life is about 5 hr (50). Because of the instability of the microtubular protein and the time required for preparation and assay, it is difficult to determine the original amount of colchicine-binding activity in the source material. To obviate this problem it has been suggested that the colchicine-binding activity be measured as a function of time after isolation and the data be extrapolated to zero time (50).

Several agents have been shown to prolong the half-life of microtubular protein as measured by its colchicine-binding activity. These include MgCl$_2$ and ATP (41); GTP (16); NaCl, sodium acetate, sodium glutamate, vincristine, and glucose (50). Colchicine at low concentrations (10^{-6} M) appears to stabilize the protein in its native state (41). However, Wilson (50) has found that high concentrations of colchicine (greater than 10^{-3}) can destroy its colchicine-binding activity, apparently through a nonspecific interaction with the protein. The effect of vinblastine on the stability of microtubular protein is unclear since several investigators report an increase in colchicine binding in the presence of vinblastine (15, 50) while others report a decrease (51).

Agents which are known to decrease colchicine-binding activity include picropodophyllotoxin, podophyllotoxin (50), and copper (52). These agents act either by decreasing the half-life of the protein or by some mode of noncompetitive inhibition.

A. Filter Assay for Colchicine Binding

A method which involves the binding of the protein-colchicine (^{3}H) complex to DEAE filter paper was first described by Weisenberg et al. (16) and is the most widely used assay method for microtubular protein.[2] It has been shown to be reliable for homogenate, soluble and resuspended particulate fractions including vinblastine precipitates (48). The original procedure along with variations and modifications suggested by other workers, is listed below.

1. Incubation Procedures

Samples are incubated in 2-ml beakers in a 37°[†] shaking water bath for 1 hr. The incubation mixture contains (final concentrations): 2.5×10^{-6} M (^{3}H)–colchicine (0.5μCi); 0.01 M sodium phosphate, pH 6.5; 0.01 M $MgCl_2$; in a final volume of 1 ml containing less than 1 mg protein. After 1 hr, the beakers are removed to an ice bath to stop the reaction. The optimal incubation temperature for colchicine is 37°; no binding occurs at 0° (50). Although 1 to 1.5 hr of incubation time is routinely used for most assay procedures, the amount of colchicine bound at this time reaches only about 85–90% of its maximal level. Maximum binding is achieved only after 2.5 hr of incubation (49).

2. Isolation of Protein–Colchicine Complex.

DEAE cellulose chromedia paper (DE81, Whatman Co.) disks, 25 mm in diameter, are equilibrated in PM buffer (0.01 M sodium phosphate, pH 6.5; 0.01 M $MgCl_2$). The disks are placed in a Millipore filter apparatus and beakers containing the samples are then emptied directly onto the disks. The beakers are rinsed with 1 ml of 10^{-4} M nonradioactive colchicine and then emptied again directly on the paper. Eight milliliters of PM buffer are added to the filter funnel and allowed to drip through under a slight vacuum. Approximately 8 min are taken for this filtration. The paper is washed by three sequential additions of 4 ml of buffer. The washing solutions are kept on ice.

3. Measurement of Colchicine

The paper disks are placed directly in counting vials containing 10 ml of counting solution (53) and counted. The radioactivity is eluted from the paper by the scintillation fluid. The presence of the paper does not appear to affect counting efficiency.

B. Modifications

Modifications of the assay procedure have been introduced by several investigators. The incubation medium has been modified by Weisenberg and Timasheff (51) to include 5 mM $MgCl_2$ rather than the 10 mM $MgCl_2$

[†] Unless otherwise stated degrees are in Centigrade.

listed in the original procedure because they found that concentrations of 10 mM $MgCl_2$ and above can cause precipitation of the protein. Wilson (50) used an incubation medium containing 20 mM sodium phosphate and 100 mM sodium glutamate (pH 6.8). Their data indicate that sodium glutamate is a better stabilizing agent than either $MgCl_2$ or GTP, and that colchicine-binding activity is stable only over a narrow pH range with maximum binding between 6.7 to 6.8.

An alternate method for washing the DE 81 chromedia paper after absorption of the protein-colchicine complex has been reported (50). Aliquots of an incubation mixture such as described above (usually 100 μl) containing free and protein-bound colchicine are applied directly to the slightly moistened paper disks at 0°. After 10 min the paper disks are washed by immersion in five successive 30-40 ml changes of 10 mM sodium phosphate buffer, pH 6.8 (5 min per wash, 0°) to remove all unbound colchicine. The paper disks with adhering protein-bound colchicine are then counted directly in a scintillation vial containing 5 ml of Bray's solution (53).

Although both DE81 paper methods are rapid and reproducible, the binding measured by these methods average about 85% of the binding as determined by the column procedure (16) described below. Separation of the protein-colchicine complex by the column method can only be used for soluble fractions, but it is considered the most accurate and the most precise method. The incubation is carried out as described above and the sample is applied to a Sephadex G100 column. The column is eluted with PM or PMG buffer and 1 ml fractions are collected and counted. For better resolution, filtration can be performed on 2.5 × 3.5 cm columns of Sephadex G200 equilibrated in PMG and run at 4° with a flow rate of 10 ml/hr regulated by means of a peristaltic pump.

A method specifically adapted for assaying colchicine-binding activity in particulate fractions was described by Feit and Barondes (35). The binding activity measured by this method is approximately the same as measured by the Weisenberg, Borisy, and Taylor filter method (54). Portions of the sample (after incubation with colchicine as described above) are carefully layered over 7 ml of 10% (w/v) sucrose in 0.01 M $MgCl_2$, 0.01 M sodium phosphate buffer (pH 7), and then centrifuged for 1 hr at 100,000 g in the Spinco 50 rotor. The supernatant fluid is then aspirated and the pellet counted. Washing and recentrifugation of the pellet does not change the results.

In earlier work by Taylor (40) colchicine binding was measured spectrophotometrically rather than radioactively. Colchicine has an absorbance maximum at 350 nm. After isolation of the protein-colchicine complex, perchloric acid is added to release the colchicine and the absorbance of the resulting supernate is measured.

III. ISOLATION PROCEDURES

Microtubular protein has been isolated from a variety of sources during the last few years. In our discussion below we have not attempted to describe every published isolation procedure but rather have selected those upon which other procedures are based and which are most commonly used. In general two types of methods have been developed for the isolation of microtubular protein. One method, described first by Weisenberg et al. (16), involves ammonium sulfate fractionation and chromatography; the second involves the specific precipitation of the protein by the vinca alkaloid, vinblastine (55). Modifications of the original procedures have been made by several groups. The basic methods, as well as various modifications, are described in detail below. The colchicine-binding property of the protein is used as the definitive method of assay.

A. Isolation of Microtubular Protein by the Weisenberg Method

In their original publication, Weisenberg et al. (16) presented two methods for the isolation and purification of the protein. The chief difference between the two is that one involves a "batch" procedure for purification on DEAE-Sephadex while the other involves gradient elution from DEAE-Sephadex.

1. Batch Procedure

Approximately 880 g of porcine brain, obtained within 1 hr of slaughter, is packed in a beaker with ice. All further operations are done at 4°. The superficial blood vessels and meninges are removed, the tissue minced with scissors, washed three times by suspension in three volumes of 0.24 M sucrose in PM buffer (0.01 M sodium phosphate, 0.01 M $MgCl_2$, pH 6.5), and strained through cheese cloth. The mince is suspended in one volume of the same solution containing in addition 10^{-4} M GTP (PMG buffer) and homogenized for 30 sec in a Sorvall Omni-Mixer at setting 9. The homogenate is centrifuged at 16,000 g for 30 min and the pellet is discarded. The supernate is brought to 32% saturation in $(NH_4)_2SO_4$ by the addition of 177 g/liter of enzyme grade $(NH_4)_2SO_4$ (Mann), added over a 15-min period. After an additional 10 min the suspension is centrifuged at 10,000 g for 20 min. The supernatant solution is retained and brought to 43% saturation by adding 71 g $(NH_4)_2SO_4$/liter of solution. After 10 min this suspension is then centrifuged at 10,000 g for 20 min. The supernate is discarded and the pellet is suspended in about 200 ml of PMG buffer by gentle homogenization with a Potter-type homogenizer. The solution is distributed into eight 50-ml conical centrifuge tubes containing 15 ml of packed DEAE-Sephadex (A-50) which has been equilibrated in PM buffer.

The contents of the tubes are stirred intermittently for 30 min to allow the protein to adsorb. After a low-speed centrifugation the DEAE-Sephadex pellet is washed twice with five volumes of 0.5 M KCl in PMG buffer, allowing 10 min per wash with intermittent stirring. (We have found that DEAE-Sephadex does not pellet well under centrifugal force and that filtration on a Buchner funnel is a more satisfactory method for collecting the resin.) The supernates are discarded. The protein is eluted from the resin by washing twice with 10 ml/wash of 0.8 M KCl in PMG buffer allowing 10 min/wash. The resin is removed by centrifugation, the supernates are combined and 248 g/liter of $(NH_4)_2SO_4$ is added to bring the solution to 43% saturation. The precipitate is collected by centrifugation at 35,000 g, dissolved in PMG buffer, and dialyzed to remove excess $(NH_4)_2SO_4$.

2. Column Procedure

In the column procedure some minor changes are made in the early steps of the purification. The minced brain is homogenized in two volumes of PM buffer instead of one, and in the $(NH_4)_2SO_4$ step, the material precipitating between 38 and 49% saturation is retained. The protein is dissolved in approximately 50 ml of PM buffer per liter of original homogenate, lyophilized, and stored at -20°.

Portions of this material are used for further purification. A 2-ml sample containing about 5-10 mg/ml of protein is applied to a 1 x 5 cm column of DEAE-Sephadex (A-50) previously equilibrated with 0.02 M sodium phosphate, 0.01 M $MgCl_2$, pH 6.5. The column is eluted first with PM buffer containing 0.1 M NaCl, then with the same buffer containing 0.3 M NaCl, and finally with an exponential gradient between 0.3 M and 0.8 M NaCl in PM buffer. Two-milliliter fractions are collected. Colchicine binding protein starts to be eluted at a NaCl concentration of about 0.5 M. A summary of the results of the purification procedure is presented in Table 1. The specific activity of the protein after the final step (1260 CPM/mg) is equivalent to the binding of about 0.4 moles of colchicine/120,000 g of protein.

An important consideration in purification procedures of labile proteins such as tubulin, is the time required. The gel elution procedure results in a protein of greater purity but requires more time. The batch procedure, on the other hand, offers a rapid method for obtaining large amounts of reasonable pure protein.

3. Modifications[3]

In a later paper (51) Weisenberg and Timasheff were able to increase the degree of purity of the protein from calf brain by introducing alterations in the "batch" procedure. In addition to slight changes in the buffers (the

TABLE 1

Purification of Colchicine Binding Protein from Porcine Brain[a]

Step	Vol, ml	Total activity,[b] x 10^{-6}	Total protein, mg	Specific activity,[c] x 10^{-3}	Yield, %
1. Homogenate	700	273	28,000	10.8	100
2. Soluble supernate	430	304	5,700	53.1	106
3. (NH$_4$)$_2$SO$_4$ (38–49%)	40	160	1,270	126	59
4. Column DEAE–Sephadex[d]		48	38	1,260	20

[a] As modified from Weisenberg et al. (16).

[b] Expressed as CPM bound.

[c] Expressed as CPM bound per milligram protein.

[d] Only a small sample of the (NH$_4$)$_2$SO$_4$ step was used. The values presented for the DEAE–Sephadex are given in terms of the total sample.

PMG buffer used contains 5 mM $MgCl_2$ and after adsorption of the protein,
the DEAE-Sephadex is washed with 0.4 M KCl instead of 0.5 M KCl) a
$MgCl_2$ precipitation step is introduced. This step was necessary because
the original procedure of Weisenberg et al. for porcine brain did not produce a pure protein when applied to calf brain. Thus original procedures
may have to be modified when used with brains from other animals. After
the last $(NH_4)_2SO_4$ step, the remaining $(NH_4)_2SO_4$ is removed by dialysis
or by Sephadex G-25 chromatography. Complete removal is apparently
essential. $MgCl_2$ (0.5 M) is added slowly to the protein solution until the
concentration of $MgCl_2$ reaches 0.05 M. After 10 min in the cold, the
precipitate which forms is collected by centrifugation and washed with
0.05 M $MgCl_2$. The final precipitate is resuspended in 0.02 M sodium
phosphate, 10^{-4} M GTP, pH 7. The protein can be lyophilized but care
must be exercised to avoid aggregation. To achieve this the authors rec-
ommend the following procedure. The protein is suspended in enough
phosphate-GTP buffer so that the concentration is 10 mg protein/ml. After
4 to 5 hr of dialysis the solution is frozen rapidly in a dry-ice-acetone
bath, lyophilized, and stored below 0°.

4. Preparation of an Akylated Protein

Falxa and Gill (18) have described a method for preparing an alkylated
protein from calf brain which is presumably the monomer (60,000 mol wt)
of microtubular protein. The procedure is based on the batch method of
Weisenberg et al. except that 10^{-2} M ascorbic acid is included in the
buffers prior to elution from DEAE-Sephadex. The protein is eluted from
DEAE-Sephadex with PMG which lacks ascorbate. After elution, iodo-
acetamide (final concentration, 0.01 M) is added to the protein. The protein is then isolated by $(NH_4)_2SO_4$ precipitation as described in the Weisenberg procedure. The disadvantage of such a preparation is the possible
alteration of the secondary and tertiary structures of the protein by the
alkylation. Physical measurements of such a derivative would not necessarily give correct information about the structure of the native protein.

5. Preparation of Microtubular Protein from Chick Embryo Brains

Weisenberg's procedure has been modified for use with chick embryo
brain which allows a one-step purification of the protein from this source
(56). The high concentration of microtubular protein in embryonic tissue
aids in the ease of purification.

Freshly dissected chick embryo brains (14 to 19 days old) are
homogenized for 30 sec in phosphate-glutamate buffer (20 mM sodium
phosphate, 100 mM sodium glutamate, pH 6.8) with a Teflon-glass homogenizer. After centrifugation at 100,000 g for 60 min the supernate,

containing about 100 mg of protein in 6 to 8 ml, is applied to a 2.5 x 6 cm
column of DEAE-Sephadex previously equilibrated with a buffer, pH 6.8,
containing 20 mM sodium phosphate and 100 mM NaCl. After the sample
is added, the column is first washed with 10 to 20 ml of the equilibrating
buffer, and then with 50 ml of 20 mM sodium phosphate, 0.4 M NaCl,
pH 6.8. The microtubular protein is then eluted using a linear gradient
of 0.4 to 0.8 M NaCl in 20 mM sodium phosphate, pH 6.8 (total volume,
250 ml). The protein elutes at a concentration of approximately 0.52 M.

B. Purification of Microtubular Protein by Precipitation with Vinblastine

The vinca alkaloid vinblastine (Fig. 5) is an antimitotic agent (57)
which apparently acts by binding to microtubular protein which makes up
the mitotic apparatus. It has also been shown to cause the formation of
intracellular crystals of microtubular protein (46, 58) and to precipitate
the protein in vitro (55). Using this property of vinblastine, Marantz et al.
introduced a method of isolating microtubular protein from brain homogenates (55) which involves the addition of the vinca alkaloid to a supernatant
fraction leading to the precipitation of microtubular protein.

In this method, pig brain is homogenized in an equal volume of 0.01 M
phosphate buffer, pH 6.5 containing 0.01 M $MgCl_2$ and 0.24 M sucrose.
The homogenate is centrifuged for 10 min at 30,000 g to remove the cell
debris. The supernate is then centrifuged at 100,000 g for 1 hr. The
high-speed supernatant solution is made 1 mM with respect to vinblastine
and the resulting precipitate is collected by centrifugation for 30 min at
100,000 g. The precipitate can be redissolved by dialysis against vin-
blastine-free buffer. (The extent to which the precipitate can be redissolved has been found to be quite variable by different investigators.) This
procedure causes the precipitation of 40 to 70% of the colchicine binding
activity of the high-speed supernate. However, by dialyzing the 100,000 g

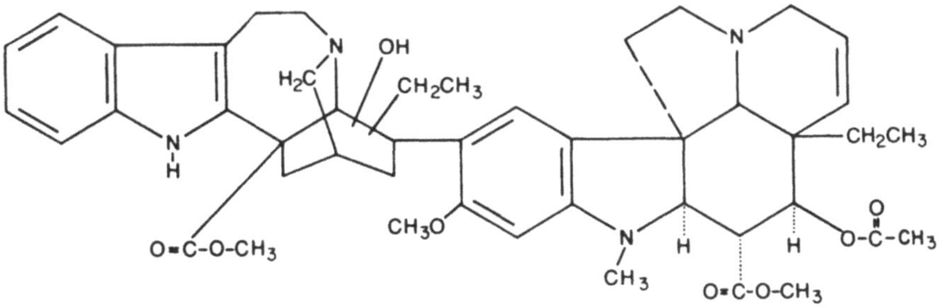

FIG. 5. Chemical formula of vinblastine.

supernate against 0.01 M phosphate buffer, pH 6.5, containing 0.01 M $MgCl_2$ for 5 hr at 4° before adding vinblastine, 99% of the colchicine binding is precipitated. Presumably components in the supernate, such as NaCl, inhibit the interaction of vinblastine and the protein. This procedure has also been applied to HeLa cells (55).

A similar procedure has been described for the microtubule protein from neuroblastoma cells (59). Cells from suspension cultures are harvested by low-speed centrifugation at room temperature and washed once with 0.01 M Tris, pH 7.0 containing 0.24 M sucrose (ST). The remaining steps are done at 4°. The cells (3×10^8 are homogenized in 3.0 ml ST or ST containing 20 mM $MgCl_2$ (SMT) and centrifuged at 1,000 g for 10 min. The supernate is retained and centrifuged at 35,000 g for 30 min. The 35,000 g supernate is used for the isolation of microtubular protein. Vinblastine is added to a final concentration of 2×10^{-3} M. If the homogenate had been prepared in ST the solution is made 2.5 mM with respect to $MgCl_2$ before the addition of vinblastine. After at least 30 min the turbid solution is centrifuged at 35,000 g for 30 min. The pellet is resuspended in 1.0 ml of SMT yielding a protein concentration of 1.5 to 2.0 mg/ml. The precipitation procedure can be repeated for further purification.

C. Isolation of Vinblastine-Induced Paracrystals

As mentioned earlier, in vivo vinblastine causes tubulin to organize into a "crystalline" structure. Nagayama and Dales (60) have used this property of vinblastine to isolate microtubular protein from L cells as a paracrystalline aggregate. To induce the formation of the paracrystals, confluent monolayer cultures of L cells are incubated for 16 hr in Eagles' (61) minimum essential Spinner medium supplemented with 10% fetal bovine serum and 10^{-5} M vinblastine (Medium A). The cells (3×10^8) in 150 ml of this medium are scraped from the plates, cooled to 4°, centrifuged at 150 g for 5 min, and resuspended in 15 ml of Spinner medium, pH 6.5, which lacks $NaHCO_3$ but contains 10^{-5} M vinblastine (Medium B). To the suspension is added 0.15 ml of Nonidet P_{40} (Shell) and the contents are agitated in a Vortex mixer for 10 sec. Fifteen milliliters of Genesolv-D (Allied Chemical) are added and the suspension is again mixed on a Vortex mixer. After centrifugation at 150 g for 3 min the upper phase is removed and retained. The lower, fluorocarbon phase contains cellular debris. The upper phase is then centrifuged at 2000 g for 15 min and the pellet, which contains the paracrystals, is resuspended in 10 ml of Medium B and the extraction procedure is repeated three times. The paracrystals isolated in this manner contain an ATPase activity (60) but the possibility that this activity is due to a contaminating protein has not been ruled out. Structurally these paracrystals resemble those produced in vitro from purified microtubule protein (62).

Vinblastine-induced crystalline bodies have also been isolated from sea urchin oocytes (63). This procedure does not require the use of a fluorocarbon phase separation to remove cell debris, as described above for L cells, and as a consequence results in an excellent yield of the crystals. Unfertilized eggs, washed three times in filtered sea water, are incubated for 18 to 36 hr with slow stirring at 13 to 14° in 10^{-4} M vinblastine. The eggs are collected by hand centrifugation and resuspended in 15 volumes of 1 M urea adjusted to pH 8.5 with 5 mM Tris-HCl. The latter step causes the lysis of the cortical granules and softens the cortex. After 2 min the cells are collected by hand centrifugation and resuspended in three volumes of 10 mM Tris-HCl, 1 mM EDTA, and 1% Nonidet P_{40} (Shell Chemical Co.), pH 7.5. Complete lysis occurs in less than 30 sec. It is possible to lose the crystals by dissolution if too much lytic buffer is added.

The crystals and some partially dispersed mitochondria are then collected by centrifugation at 1500-2000 g for 15 min. The supernate is removed by aspiration and the pellet is resuspended in a stabilizing medium containing 100 mM KCl, 1 mM $MgCl_2$, 0.1% Nonidet, 10 mM Tris-HCl, pH 7.5, and 10^{-5} M vinblastine. If the crystals are resuspended in the lysis buffer, complete dispersion occurs. Two washings of the crystals with the stabilizing medium are done to further purify the crystals. A recent study of the biochemical properties of paracrystals isolated from sea urchin eggs indicate that their colchicine binding activity, GTP content, vinblastine binding, electrophoretic mobility, and amino acid composition are identical to microtubule protein (64).

D. Isolation of Intact Microtubules

Methods have been published for the isolation of intact microtubules in an enriched but not pure state. By using hexylene glycol as a stabilizing agent, Kirkpatrick et al. (65) prepared an enriched fraction of microtubules from rat brain. A 16.7% homogenate of whole brain is made in medium containing 1 M hexylene glycol and 20 mM potassium phosphate, pH 6.4, at 1°, using a 1-in. diameter Teflon in glass homogenizer. The homogenate is centrifuged for 30 min at 48,000 g and the supernate is retained. Further purification is achieved by discontinuous sucrose gradient centrifugation. Sucrose solutions with densities of 1.16 g/ml and 1.19 g/ml are made in the buffered hexylene glycol medium and 1 ml each is layered in 4-ml plastic centrifuge tubes. Two ml of the 48,000 g supernate is added and centrifugation is carried out in a swinging bucket rotor at 39,000 rpm for 1 hr. After removal of the supernatant fraction the interface fraction between the 1.16 and 1.19 density layers is collected from a hole placed in the side of the tube 1 cm from the bottom. This fraction contains microtubules and other material as detected by electron microscopy. Kirkpatrick et al. (65) reported a purification of 122-fold based on the number of microtubules per amount of protein with an 8% recovery of the tubules seen in the original homogenate. This method may be selective for the more stable types of microtubules and probably includes microtubules from both nonneuronal cells as well as neuronal.

Preparations of axoplasm which are rich in microtubules have been obtained by simply extruding the axoplasm from the giant axons of the squid Dosidicus gigas (66). It was found that stabilization of the micro-tubules from this source was best achieved with a D_2O-H_2O solution (>40% D_2O) of low ionic strength lacking divalent cations. Sodium citrate or EDTA, both 0.01 M, at pH values between 6 and 6.5 in D_2O were effective. In this system the authors found that hexylene glycol was ineffective in preserving the microtubular structure.

E. Summary

The procedures described above represent the commonly used methods for the isolation of microtubular protein. In addition there are published procedures for the isolation of microtubular protein from sperm tail (67), Chlamydomonas flagella (15), and Tetrahymena cilia (23). Basically the Weisenberg procedure, or variations of it, and the vinblastine precipita-tion method represent the two methods available for the purification of microtubular protein from brain. One of the major difficulties for workers in this field is the lability of the protein as detected by the rapid loss of colchicine binding. Some stabilization of the protein has been achieved (see page 122) but even in the presence of stabilizing agents rather rapid inactivation occurs. Because of this instability a fast purification proce-dure is desired. The vinblastine precipitation method does fulfill this requirement and has the added advantage that the protein can be isolated from small samples of tissue. However, there is considerable variability in the purity of the protein isolated by vinblastine precipitation from dif-ferent tissues. This is in part due to the influence of other factors present in extracts of tissues but largely due to the lack of strict specificity of vinblastine. It has been shown, for example, that under certain conditions vinblastine will precipitate a number of proteins as well as DNA and ribo-somes (68). Moreover, very little is known about the binding of vinblastine to the protein. Apparently one mole of vinblastine binds per dimer (64); and probably at the exchangeable GTP binding site (69) but little is known about the structural alterations which occur in the protein. Vinblastine does induce irreversible changes in the structure of the protein (51) since the molecular weight of the alkaloid precipitated protein, even after exten-sive dialysis, is over 250,000 (59). Because of these uncertainties with the vinblastine method, the Weisenberg procedure is the preferred isola-tion procedure if one wishes to study physical and chemical properties of microtubular protein.

IV. PROPERTIES AND PURITY

A. Heterogeneity

As a general rule, the physical and chemical properties of any particular protein will vary according to the source material from which it is
isolated. In the case of hemoglobins this has been important in evolution-
ary considerations, diseased states, etc.; with isoenzymes, these variations are related to the properties of the polymerized forms. Although a
striking feature of tubulin is the similarity of its properties regardless of
source, some small but significant diversity among tubulins from different
sources have been reported. Stephens (67) demonstrated a dissimilarity
in aminoacid composition and peptide maps of tubulin from A-tubules and
B-tubules obtained by thermal fractionation of flagellar outer doublet fibers.
The name A-tubulin was proposed for protein from A-tubules and B-tubulin
for protein from B-tubules.

Refinements in the preparation and characterization of tubulin have
revealed an additional dissimilarity which investigators have termed
"heterogeneity." When tubulin dimers, isolated from a single source, are
denatured in urea or SDS, two types of monomers (called α and β) are distinguishable. The nomenclature used for the designation of these two monomer types, differs from author to author. We suggest that Bryan's nomenclature (α) and (β) be used to denote the two types of monomers commonly
found in most preparations (56).

Two tubulin monomers which are distinguishable on the basis of electrophoretic mobility and amino acid composition are present in chick
embryo brains (56) and vinblastine-induced crystalloids from sea urchin
eggs (64). However, no differences were found in the molecular weight
of the monomers (56, 64). Olmsted et al. (15) isolated two tubulin monomers from neuroblastoma cells and porcine brain which differed in molecular weights (56,000 and 53,000 respectively) as well as aminoacid
composition. Feit et al. (14) also found monomers with molecular weights
of 56,000 and 53,000 in adult pork and mouse brain, and in neuroblastoma
cells. The monomers exhibited different peptide maps and isoelectric
points. [4]

These data do not necessarily indicate that a given tubule is composed
of two types of monomers. It is possible that microtubular preparations
are composed of a mixture of homotubules each containing only one type
of protein. However, Olmsted's (15) finding that α and β monomers are
present in preparations containing only A-tubules and in preparations containing only B-tubules, suggests that tubules are indeed heteropolymers,
i.e., they are made up of two different types of monomers.

Some species differences between tubulin dimers appear to be a result of differences between the β monomers only. The α monomers from three sources, neuroblastoma, brain, and flagella, appear to be identical electrophoretically (15). However, the β monomer from flagella has a different electrophoretic mobility than does the monomer from neuroblastoma and brain (15).

Additional monomer types have been identified in immature brain. Isoelectric focusing of tubulin from immature mouse brain showed a splitting of each of the two bands representing the α and β monomers into two finer bands (14). Four types of tubulin monomers were also found in cultured chick embryo cerebral ganglia, with molecular weights ranging from 50,000-70,000 D (70).

In summary, tubulin from most sources appears to contain a minimum of two different kinds of monomers (α and β) but the arrangement of these monomers in the microtubules is not known. However, since both kinds of tubulin monomers are found in the same kind of microtubule (e.g., both are found in A-and B-tubules) and in approximately equal amounts (14, 15, 64) it seems likely that the two types of monomers join to form heterodimers which bind colchicine and vinblastine, and presumably form the polymerizing unit for microtubules.

An important question is whether or not these slight but perhaps significant differences in tubulin monomers can be related to the properties or functions of microtubules. For example, does the extent of heterogeniety have a bearing on the function or the stability of a particular type of tubule (e.g., neurotubules, axial unit A and B tubules) ?

B. Chemical Composition

The amino acid composition of microtubular protein from a wide variety of sources has been analyzed and shows little variation (Table 2, Fig. 6). However there are some differences between the α and β monomers from the same source (e.g., chick embryo brain, Table 2).[5] The cysteine content is nearly identical in all sources measured even though the methods of analysis include simple acid hydrolysis, performic acid oxidation, titration with 5,5'-dithiobis-(2-nitrobenzoic acid), and analysis after carboxymethylation. This would suggest the absence of disulfide bonds. However Stephens (67) has reported that the B-tubulin from the outer fiber doublet of sea urchin sperm contains one disulfide bond.

Microtubular protein contains two GTP binding sites per dimer, a tight one and a loose one (16). Upon isolation of the protein, some guanine nucleotides remain bound. As far as we are aware, there do not appear to be any extensive studies on the possible presence of metals, or lipid in

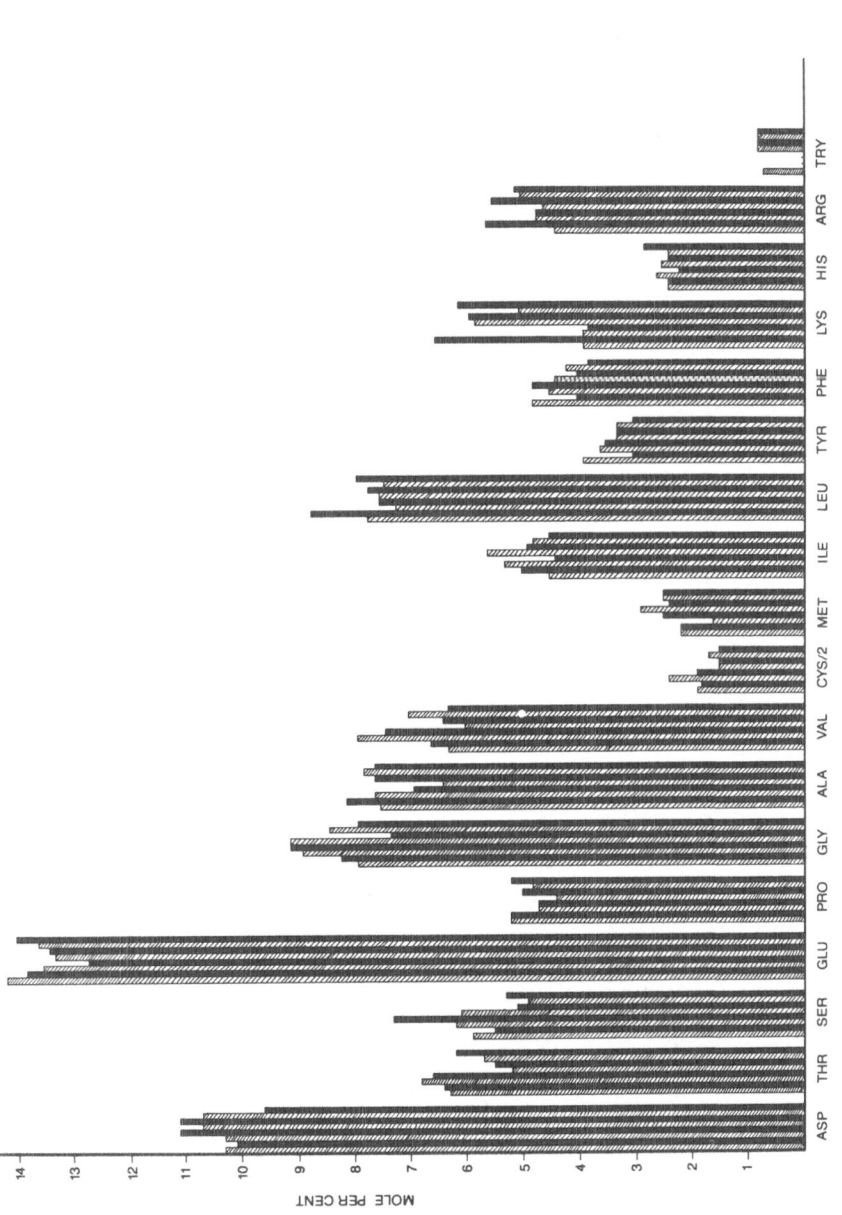

FIG. 6. Amino acid composition of microtubule protein. The bars represent tubulin from different sources. Reading from left to right the sources are: pork brain, neuroblastoma, chick embryo brain α band, chick embryo brain β band, cilia outer fiber, sea urchin sperm outer fiber A-tubulin, sea urchin sperm outer fiber B-tubulin, sea urchin sperm tail central pair.

TABLE 2

Amino Acid Composition of Microtubular Protein[a]

Residue	Pork Brain (16)	Neuroblastoma (59)	Chick embryo brain (56)		Cilia outer fiber[f] (71)	Sea-urchin sperm outer fiber[f] (67)		Sea-urchin sperm central pair (72)
			α	β		A-tubulin	B-tubulin	
ASP	10.3	10.1	10.3	11.1	10.7	11.1	10.7	9.6
THR	6.3	6.4	6.8	6.6	5.2	5.5e	5.7e	6.2
SER	5.9	5.5	6.2	7.3	6.1	5.1e	4.9e	5.3
GLU	14.2	13.8	13.5	12.7	13.3	13.4	13.6	14.0
PRO	5.2	5.2	4.7	4.7	4.4	5.0	4.8	5.2
GLY	7.9	8.2	8.9	9.1	9.1	7.3	8.4	7.9
ALA	7.5	8.1	7.6	6.9	6.3	7.6	7.8	7.6
VAL	6.3	6.6	7.9	7.4	6.0	6.4	7.0	6.3
CYS/2	1.9	1.8	2.4b	1.9b	1.5	1.5g	1.7g	1.5d
MET	2.3	2.3	1.6	2.5	2.9	2.4h	2.5h	2.5
ILE	4.5	5.0	5.3	4.4	5.6	4.9	4.8	4.5
LEU	7.7	8.7	7.2	7.5	7.5	7.7	7.4	7.9
TYR	3.9	3.0	3.6	3.5	3.3	3.3c	3.3c	3.0
PHE	4.8	4.0	4.5	4.8	4.4	4.0	4.2	3.8

LYS	3.9	6.5	3.9	3.8	5.8	5.9	5.0	6.1
HIS	2.4	2.4	2.6	2.2	2.5	2.4	2.4	2.8
ARG	4.4	5.6	4.7	4.7	4.6	5.5	5.0	5.1
TRY	0.7	–	–	–	0.8c	0.8c	0.8c	0.8c

[a] Expressed as mole %.

[b] Determined as carboxymethyl-cysteine.

[c] Determined spectrophotometrically in base.

[d] Determined by titration in 8 M urea with 5,5'-dithiobis-(2-nitrobenzioc acid).

[e] Determined by extrapolation to zero hydrolysis time.

[f] Calculated from data reported in refs.

[g] Determined as cysteic acid after performic acid oxidation.

[h] Determined as methionine sulfone after performic acid oxidation.

tubulin. One report (18) indicates that the carboxymethylated protein from
beef brain (mol wt 70,000) contains 1% carbohydrate.[1] However since
the protein was purified using Sephadex (a dextran) it is difficult to know
if the carbohydrate is a part of the protein or is a contaminate from the
Sephadex.

C. Molecular Weight

The accepted molecular weight for the colchicine binding form of
microtubular protein is 110,000-120,000 D. This protein is made up of
two monomeric units of 55,000-60,000 D. One of the major problems in
obtaining good sedimentation data has been the tendency of the protein to
aggregate. By using the Yphantis meniscus depletion method which allows
for the use of low concentrations of protein and hence helps avoid aggregation, Shelanski and Taylor (72) found that tubulin preparations from central-
pair microtubules from sea-urchin sperm tail contained two species of
protein, one with a molecular weight of 120,000 D and another with a molecular weight of 60,000 D. The 120,000 unit has an $S°_{20,W}$ value of 6.2;
however, this value is greatly dependent on the Mg^{2+} concentration (51).
In the presence of 6 M guanidine hydrochloride and 2-mercaptoethanol,
only the 60,000 D unit was observed. Essentially identical results were
observed with the outer-doublet microtubules. Moreover by Sephadex
G-200 chromatography it was shown that colchicine binds to the 120,000
but not the 60,000 mol wt protein. After a two-week exposure to 6 M guanidine hydrochloride a 30,000 D protein was observed (72). Because this
species is not usually seen in microtubular protein and occurs only after
an extended exposure to 6 M guanidine hydrochloride, it is probable that
it arose because of some nonspecific hydrolysis. A similar investigation
of the microtubular protein from beef brain (16) resulted in identical values
for the molecular weights of the dimeric and monomeric units. More
recently, SDS-gel electrophoresis has been used to confirm the dimeric
nature of tubulin and to show the nonidentical nature of the subunits (14,
15, 64). As described above, the α and β monomeric units of the protein
from neuroblastoma cells, flagella, or pork brain differ slightly in molecular weight as well as in amino acid composition.

D. The Presence of Enzymic Activities

Several laboratories have reported that microtubular protein has
certain enzymic and substrate activities all of which are associated with
phosphorylation by high-energy nucleoside phosphates. For example,
Soifer (73) found that porcine brain tubulin contains a protein kinase activity capable of catalyzing the phosphorylation of casein by ATP. Berry
et al. (74) have shown that under certain conditions GTP binding at the

exchangeable site in tubulin is accompanied by removal of the γ phosphate and phosphorylation of GDP at the nonexchangeable site. ATPase activity of microtubular protein isolated in vivo from L cells by vinblastine-induced paracrystal formation has also been reported (60). Goodman et al. (75) have shown that bovine brain microtubular protein can be phosphorylated by ATP in a reaction catalyzed by an "intrinsic" protein-kinase found in association with the tubulin stimulated by cyclic-AMP. Similarly Murry and Froscio (76) found that rat brain could be phosphorylated by a cyclic AMP-dependent protein kinase from hog brain. On the other hand, tubulin isolated from sea urchin eggs as vinblastine-induced crystals contained no ATPase or cyclic-AMP dependent protein kinase activity (64).

Whether the reported enzymic activities associated with tubulin are actual properties of the protein and perhaps related to its function or are simply artifacts such as the presence of small amounts of contaminating proteins still remains to be clarified.

V. CONCLUSIONS

Three major advances within the past decade have contributed enormously to the study of microtubular systems. These are:

(1) Improved methods of tissue fixation for electron microscopy, in particular the use of glutaraldehyde which gives a better preservation of microtubules. As a consequence, microtubules have been established as an organelle present in all eukaryotic cells.

(2) The finding that tubulin has a unique affinity for colchicine. This interaction is the basis for the assay for tubulin.

(3) The finding that tubulin interacts with and is precipitated by the antitumor agent vinblastine.

These advances have stimulated interest in microtubular systems by providing a multidisciplinary approach to their distribution, dynamics morphology, chemistry, and function. In addition the possibility has emerged that tubulin may be present as an integral part of some membranes. Since microtubules and tubulin are implicated in many translocation (e.g., axoplasmic transport, release of secretory products), and chemo-mechanical events (e.g., flagella movement), research on these components promises to open up new possibilities in understanding related events in the nervous system.

Perhaps the most intriguing yet speculative concept about the microtubules is their potential as an intracellular communication system. The common presence of microtubules in sensory transduction structures, the linear arrangements of microtubules along sizable distances in axons, and

the possibility of a conformational change being conducted via the thousands
of subunits in a microtubule make them candidates for intracellular communication. Transmission of information between the distant parts of the
neurons might be achieved not only by membrane phenomena such as the
action potential but also by microtubular systems.

ACKNOWLEDGMENTS

The research in this laboratory related to microtubules and tubulin
was supported by a grant from the National Institute of Neurological
Diseases and Stroke, NINDS 01151 and a grant from the University of
Kansas Biomedical Sciences Research fund. Richard H. Himes is an NIH
Career Development Awardee, GM 13320.

REFERENCES

1. K. R. Porter, "Cytoplasmic microtubules and their functions," in Ciba
   Foundation Symposium on Principles of Biomolecular Organization
   (G. E. W. Wolstenholme and M. O'Connor, eds.), Little, Brown,
   Boston, 1966, p. 308.

2. F. O. Schmitt and F. E. Samson, Jr., Neuronal fibrous proteins,
   Neurosci. Res. Progr. Bull., 6, 113 (1968).

3. R. E. Hinkley, Jr., and L. S. Green, Effects of halothane and colchicine on microtubules and electrical activity of rabbit vagus nerves,
   J. Neurobiol., 2, 97 (1971).

4. A. Peters, S. L. Palay, and H. de F. Webster, The Fine Structure
   of the Nervous System, Harper and Row, New York, 1970, p. 58-64.

5. A. Peters and J. E. Vaughn, Microtubules and filaments in the axons
   and astrocytes of early postnatal rat optic nerves, J. Cell Biol., 32,
   113 (1967).

6. H. Morhi, Amino acid composition of "tubulin" constituting microtubules of sperm flagella, Nature, 217, 1053 (1968).

7. L. G. Tilney, Microtubules in the asymmetric arms of Actinosphaerium
   and their response to cold, colchicine and hydrostatic pressure, Anat.
   Rev., 15, 426 (1965).

8. D. Marsland, The action of hydrostatic pressure on cell division,
   Ann. N. Y. Acad. Sci., 51, 1327 (1951).

9. D. Marsland, Cell division--enhancement of the anti-mitotic effects of
   colchicine by low temperature and high pressure in the cleaving eggs
   of Lytechinus variegatus, Exptl. Cell Res., 50, 369 (1968).

10. D. Marsland and A. M. Zimmerman, Structural stabilization of the mitotic apparatus by heavy water in the cleaving eggs of Arbacia punctulata, Exptl. Cell. Res., 38, 306 (1965).

11. S. Inoue and H. Sato, Cell motility by labile association of molecules. The nature of mitotic spindle fibers and their role in chromosome movement, J. Gen. Physiol., 50, 259 (1967).

12. H. J. Arnott and H. E. Smith, Analysis of microtubule structure in Euglena granulata, J. Phycal., 5, 68 (1969).

13. G. G. Borisy and E. W. Taylor, The mechanism of action of colchicine. Binding of colchicine -^{3}H to cellular protein, J. Cell Biol., 34, 525 (1967).

14. H. Fcit, L. Slusarek, and M. L. Shelanski, Heterogeneity of tubulin subunits, Proc. Natl. Acad. Sci. USA, 68, 2028 (1971).

15. J. B. Olmsted, G. B. Witman, K. Carlson, and J. L. Rosenbaum, Comparison of the microtubule proteins of neuroblastoma cells, brain, and Chlamydomonas flagella, Proc. Natl. Acad. Sci. USA, 68, 2273 (1971).

16. R. C. Weisenberg, G. G. Borisy, and E. W. Taylor, The colchicine-binding protein of mammalian brain and its relation to microtubules, Biochem., 7, 4466 (1968).

17. P. R. Burton, Optical diffraction and translational reinforcement of microtubules having a prominent helical wall structure, J. Cell Biol., 44, 693 (1970).

18. M. L. Falxa and T. J. Gill, Preparation and properties of an alkylated brain protein related to the structural subunit of microtubules, Arch. Biochem. Biophys., 135, 194 (1969).

19. E. Tani and T. Ametani, Substructure of microtubules in brain nerve cells as revealed by ruthenium red, J. Cell Biol., 46, 159 (1970).

20. N. J. Lane and J. E. Treherne, Lanthanum staining of neurotubules in axons from cockroach ganglia, J. Cell Sci., 7, 217 (1970).

21. P. R. Burton and H. L. Fernandez, Filamentous material associated with the surfaces of axonal microtubules, Am. Soc. Cell Biol. Abstr., Nov. 17-20, 1971, p. 41.

22. D. S. Smith, On the significance of cross-bridges between microtubules and synaptic vessels, Phil. Trans. Roy. Soc. London B., 261, 395 (1971).

23. I. R. Gibbons, Chemical dissection of cilia, Arch. Biol. (Liège), 76, 317 (1965).

24. L. E. Roth, D. J. Pihlaja, and Y. Shigenaka, Microtubules in the heliozoan axopodium, J. Ultrastruct. Res., 30, 7 (1970).

25. M. A. Rudzinska, Ultrastructures involved in the feeding mechanism of Suctoria, Trans. N. Y. Acad. Sci., 29, 512 (1967).

26. O. Behnke and A. Forer, Evidence for four classes of microtubules in individual cells, J. Cell Sci., 2, 169 (1967).

27. R. E. Hinkley, Temperature and drug induced alterations of axonal microtubules, University of Kansas, Ph.D. thesis, Lawrence, Kansas, 1971.

28. D. T. Moran and F. G. Varela, Microtubules and sensory transduction, Proc. Natl. Acad. Sci. U.S.A., 68, 757 (1971).

29. L. G. Tilney and K. R. Porter, Studies on the microtubules in heliozoa. II. The effect of low temperature on these structures in the formation and maintenance of the axopodia, J. Cell Biol., 34, 327 (1967).

30. O. Behnke, "Microtubules in Disk-Shaped Blood Cells," in International Review of Experimental Pathology, Vol. 9, Academic, New York & London, 1970, pp. 1-92.

31. H. L. Fernandez, P. R. Burton, and F. E. Samson, Axoplasmic transport in the crayfish nerve cord. The role of fibrillar constituents of neurons, J. Cell Biol., 51, 176 (1971).

32. F. O. Schmitt, Fibrous proteins—neuronal organelles, Proc. Natl. Acad. Sci. U.S.A., 60, 1092 (1968).

33. S. E. Malawista, The melanocyte model, J. Cell Biol., 49, 848 (1971).

34. D. Bikle, L. G. Tilney, and K. R. Porter, Microtubules and pigment migration in the melanophores of Fundulus heteroclitus L., Protoplasma, 61, 322 (1966).

35. H. Feit and S. Barondes, Colchicine binding activity in particulate fractions of mouse brain, J. Neurochem., 17, 1355 (1970).

36. S. L. Twomey and F. E. Samson, Tubulin antigenicity in brain particulates, Brain Res., 37, 101 (1972).

37. N. B. Thoa, G. F. Wooten, J. Axelrod, and I. J. Kopin, Inhibition of release of dopamine-β-hydroxylase and norepinephrine from sympathetic nerves by colchicine, vinblastine, or cytochalasin-B, Proc. Natl. Acad. Sci. U.S.A., 69, 520 (1972).

38. A. M. Poisner and J. Bernstein, A possible role of microtubules in catecholamine release from the adrenal medulla: effect of colchicine, vinca alkaloids, and deuterium oxide, J. Pharmacol. Exptl. Therap., 171, 102 (1971).

39. L. S. Goodman and A. Gilman, The Pharmacological Basis of Therapeutics, MacMillan, New York, 1955, p. 304-307.

40. E. W. Taylor, The mechanism of colchicine inhibition of mitosis,
J. Cell Biol., 25, 145 (1965).

41. G. G. Borisy and E. W. Taylor, The mechanism of action of colchicine, J. Cell Biol., 34, 535 (1967).

42. M. L. Shelanski and E. W. Taylor, Isolation of a protein subunit from
microtubules, J. Cell Biol., 34, 549 (1967).

43. O. J. Eigsti and P. Dustin, Colchicine in Agriculture, Medicine,
Biology and Chemistry, State College Press, Ames, Iowa, 1955.

44. S. S. Malawista, The action of colchicine in acute gout, Arthritis
Rheumat., 8, 752 (1965).

45. E. Robbins and N. K. Gonatas, Histochemical and ultrastructural
studies on HeLa cell cultures exposed to spindle inhibitors with special
reference to the interphase cell, J. Histochem. Cytochem., 12, 704
(1964).

46. H. Wisniewski, M. L. Shelanski, and R. D. Terry, Effects of mitotic
spindle inhibitors on neurotubules and neurofilaments in anterior horn
cells, J. Cell Biol., 38, 224 (1968).

47. W. A. Creasy and M. E. Markiw, Biochemical effects of the vinca
alkaloids. II. A comparison of the effects of colchicine, vinblastine,
and vincristine on the synthesis of ribonucleic acids in Ehrlich ascites
carcinoma cells, Biochim. Biophys. Acta, 87, 601 (1964).

48. D. R. Dahl, D. A. Redburn, and F. E. Samson, Jr., Regional distribution of colchicine-binding (microtubular) protein in the rat brain,
J. Neurochem., 17, 1215 (1970).

49. L. Wilson and M. Friedkin, The biochemical events of mitotis. II.
The in vivo and in vitro binding of colchicine in grasshopper embryos
and its possible relation to inhibition of mitosis, Biochemistry, 6,
3126 (1967).

50. L. Wilson, Properties of colchicine binding protein from chick embryo
brain. Interactions with vinca alkaloids and podophyllotoxin, Biochemistry, 9, 4999 (1970).

51. R. C. Weisenberg and S. N. Timasheff, Aggregation of microtubule
subunit protein. Effects of divalent cations, colchicine and vinblastine,
Biochemistry, 9, 4110 (1970).

52. L. Wilson and J. Bryan, Copper (II) interaction with microtubule protein of chick embryo brain, Federation Proc., 29, 941abs. (1970).

53. G. A. Bray, A simple efficient liquid scintillator for counting aqueous solutions in a liquid scintillation counter, Anal. Biochem., 1, 279 (1960).

54. D. A. Redburn, Neuronal microtubules, microtubular protein and neurolathyrism, University of Kansas, Ph. D. thesis, Lawrence, Kansas, 1972.

55. R. Marantz, M. Ventilla, and M. L. Shelanski, Vinblastine-induced precipitation of microtubule protein, Science, 165, 498 (1969).

56. J. Bryan and L. Wilson, Are cytoplasmic microtubules heteropolymers? Proc. Natl. Acad. Sci. U.S.A., 68, 1762 (1971).

57. P. George, L. J. Journey, and M. N. Goldstein, Effect of vincristine on the fine structure of HeLa cells during mitosis, J. Natl. Cancer Inst., 35, 355 (1965).

58. K. G. Bensch and S. E. Malawista, Microtubule crystals: a new biophysical phenomenon induced by vinca alkaloids, Nature (London), 218, 1176 (1968).

59. J. B. Olmsted, K. Carlson, R. Klebe, F. Ruddle, and J. Rosenbaum, Isolation of microtubule protein from cultured mouse neuroblastoma cells, Proc. Natl. Acad. Sci. U.S.A., 65, 129 (1970).

60. A. Nagayama and S. Dales, Rapid purification and the immunological specificity of mammalian microtubular paracrystals possessing an ATPase activity, Proc. Natl. Acad. Sci. U.S.A., 66, 464 (1970).

61. H. Eagle, Nutrition needs of mammalian cells in tissue culture, Science, 122, 501 (1955).

62. R. Marantz and M. L. Shelanski, Structure of microtubular crystals induced by vinblastine in vitro, J. Cell Biol., 44, 234 (1970).

63. J. Bryan, Vinblastine and microtubules. I. Induction and isolation of crystals from sea urchin oocytes, Exptl. Cell Res., 66, 129 (1971).

64. J. Bryan, Biochemical characteristics of isolated vinblastine induced crystals, Am. Soc. Cell Biol. Abst., Nov. 17-20, 1971, p. 38.

65. J. B. Kirkpatrick, L. Hyams, V. L. Thomas, and P. M. Howley, Purification of intact microtubules from brain, J. Cell Biol., 47, 384 (1970).

66. P. F. Davison and F. C. Huneeus, Fibrillar proteins from squid axons II. Microtubule protein, J. Mol. Biol., 52, 429 (1970).

67. R. E. Stephens, Thermal fractionation of outer fiber doublet microtubules into A-and-B-subfiber components: A-and-B-tubulin, J. Mol. Biol., 47, 353 (1970).

68. L. Wilson, J. Bryan, A. Ruby, and D. Mazia, Precipitation of proteins by vinblastine and calcium ions, Proc. Natl. Acad. Sci. U.S.A., 66, 807 (1970).

69. M. Ventilla, C. R. Cantor, and M. L. Shelanski, Vinblastine effect on GTP binding to microtubule protein, Federation Proc., 29, 290abs. (1970).

70. R. N. Bryan, Colchicine-binding proteins of cultured brain cells, Trans. Am. Soc. Neurochem., 2, 59 (1971).

71. F. L. Renaud, A. J. Rowe, and I. R. Gibbons, Some properties of the protein forming the outer fibers of cilia, J. Cell Biol., 36, 79 (1968).

72. M. L. Shelanski and E. W. Taylor, Properties of the protein subunit of central-pair and outer-doublet microtubules of sea urchin flagella, J. Cell Biol., 38, 304 (1968).

73. D. Soifer, Intrinsic protein kinase activity of tubulin from porcine brain, Am. Soc. Cell Biol. Abst., Nov. 17–20, 1971, p. 283.

74. R. W. Berry, M. Ventilla, C. Cantor, and M. L. Shelanski, Microtubule protein: vinblastine and GTP interactions, Am. Soc. Cell Biol. Abst., Nov. 17–20, 1971, p. 29.

75. D. B. P. Goodman, H. Rasmussen, F. DiBella, and C. E. Guthrow, Jr., Cyclic adenosine 3':5'-monophosphate-stimulated phosphorylation of isolated neurotubule subunits, Proc. Natl. Acad. Sci. U.S.A., 67, 652 (1970).

76. A. W. Murray and M. Froscio, Cyclic adenosine 3':5'-monophosphate and microtubule function: specific interaction of the phosphorylated protein subunits with a soluble brain component, Biochem. Biophys. Res. Comm., 44, 1089 (1971).

77. R. K. Margolis, R. U. Margolis, and M. L. Shelanski, The carbohydrate composition of brain microtubule protein, Biochem. Biophys. Res. Comm., 47, 432 (1972).

78. B. A. Eipper, Rat brain microtubule protein: Purification and determination of covalently bound phosphate and carbohydrate, Proc. Natl. Acad. Sci. U.S.A., 69, 2283 (1972).

79. G. G. Borisy, A rapid method for quantitative determination of microtubule protein using DEAE-cellulose filters, Anal. Biochem., 50, 373 (1972).

80. R. J. Owellen, A. H. Owens, Jr., and D. W. Donigian, The binding of vincristine, vinblastine and colchicine to tubulin, Biochem. Biophys. Res. Comm., 47, 685 (1972).

NOTES ADDED IN PROOF

(1) Two conflicting reports concerning the carbohydrate content of brain tubulin have recently appeared. Margolis et al. (77) claim the presence of amino and neutral sugars covalently linked to the tubulin. However, Eipper (78) found that tubulin prepared by the Weisenberg method is contaminated by nucleic acid and carbohydrate. By using a modified procedure Eipper found that tubulin contains no covalently bound amino sugars and no more than 0.2% neutral sugar.

(2) A thorough study of the important conditions for the assay has recently been published by Borisy (79).

(3) Other recent modifications in the isolation procedure are found in (78) and (80).

(4) Eipper (78) found that alkylated tubulin from rat brain migrates as two bands in SDS-urea gels but only as one in SDS gels. She suggests that in the commonly used SDS-urea gels the mobility of the protein is a function of charge as well as size.

(5) Differences in the amino acid composition between the α and β chains have also been reported for rat brain tubulin (78).

Chapter 4

ION TRANSPORT IN THE SYNAPTOSOME AND Na^+-K^+-ATPase

Ata A. Abdel-Latif
Department of Cell and Molecular Biology
Medical College of Georgia
Augusta, Georgia

I. INTRODUCTION

The availability of techniques for the preparation of synaptosomes from brain homogenates has made possible the morpho-biochemical and neuropharmacological investigation of this functionally important part of the neuron in vitro. Reports from several laboratories have shown that the synaptosomal cytoplasm can carry on a host of metabolic activities (for reviews see Refs. 1-5) which appear to be similar to those of the perikaryon of the neuron. The constituents of the synaptosome, which include the intraneuronal cytoplasm, mitochondria, and synaptic vesicles, are also separated from the outside environment by a permeability barrier with properties characteristic of a neuronal membrane. Recently there has been a crescendo of research on transport Na^+-K^+-ATPase in the synaptosome, the transport of ions across its limiting membrane, and the effect of various metabolic inhibitirs and pharmacological agents thereon (6).

A considerable amount of evidence has accumulated in the past few years showing the Na^+-K^+-ATPase system, first described by Skou (7) in the crab nerve, to be involved in the active transport of Na^+ and K^+ across the cell membrane. Studies on the subcellular distribution of this enzyme system in nervous tissue showed the synaptosome to be enriched in Na^+-K^+-ATPase, which appears to reside in the synaptosomal membrane (8-16). The transport processes across the limiting membrane of synaptosomes have been studied in the past by several investigators; and these processes are well established for the uptake of choline (17-20); sodium (21); potassium (22); sodium, potassium, and chloride (23); tryptophan (24); norepinephrine (16, 25-27); and calcium (28). Since the synaptosome can be regarded as a nonnucleated cell, retaining many of the constituents of the cell cytoplasm including the metabolic processes of intact cells, studies on its Na^+-K^+-ATPase and transport of ions across its limiting membrane should provide information relevant to: (a) its various metabolic activities; (b) function in synaptic transmission; (c) the properties of its limiting membrane and the similarities and dissimilarities with other cell membranes; and (d) the mechanism of the affect of metabolic inhibitors and drugs on the functional activity of the synapse.

Detailed description of techniques used for isolation of synaptosomes, assay for synaptosomal Na^+-K^+-ATPase, and movement of ions across the synaptosomal membrane is already available in the literature (1-5), a number of techniques used in the author's laboratory are discussed in this review.

II. PREPARATION OF SYNAPTOSOMES

A. Principle

The brain tissue is disrupted by homogenization in 0.25 M sucrose, pH 7.4. The crude mitochondrial fraction is separated from the cell debris, red blood cells, nuclei, and soluble fraction by means of differential centrifugation. Synaptosomes, free mitochondria, and myelin are then separated according to their equilibrium densities by means of density-gradient centrifugation in a sucrose or Ficoll-sucrose medium.

B. Reagents and Equipment

Sucrose, 0.25 M, pH 7.4 (adjusted with tris buffer) and containing 10^{-4}M ethylenediaminetetraacetic acid (EDTA).

Ficoll (purified), 20%, 16%, 12%, 8%, and 2% in 0.25 M sucrose. The Ficoll, which was purchased from Pharmacia Fine Chemicals, Inc., is first dialyzed for three days against deionized water to remove the sodium chloride and then lyophilized.

Centrifuge: Spinco L_2-65 B ultracentrifuge with rotor types 40 and SW 25.1 or SW 25.2.

Homogenizer: Potter-Elvehjem homogenizer with teflon pestle and 0.26 mm clearance.

Stirring motor: With low speed shaft 800 rpm (Tri-R Instrument, Inc.).

Density-gradient withdrawing device: See Fig. 3 for this easy to construct device.

C. Preparation of Brain Homogenate

Charles River Sprague Dawley derived rats (males and females), ranging in age from 12 days to adult, are stunned, decapitated, and the whole brains removed to ice-cold 0.25 M sucrose. The brains are washed twice with sucrose, and the blood vessels removed. The cerebra, cortices, or whole brains are then homogenized in sucrose. All subsequent operations are carried out at 0°. In general about 8 g of brain tissue are homogenized in 10 vol of 0.25 M sucrose by 12 strokes (up and down) at 800 rpm of a loose-fitting (0.25 mm clearance) Teflon pestle-glass homogenizer.

The tube is immersed in ice during this procedure. Homogenization should be discontinued when brain tissue is no longer discernible. Excessive homogenization damages both the mitochondria and synaptosomes and should be avoided.

D. Isolation of Synaptosomes

The homogenate is subfractionated as outlined in Fig. 1. First it is distributed into cellulose nitrate tubes, 5/8 in. diameter x 3 in. and centrifuged at 1000 g for 10 min. The sediment (crude mitochondrial fraction) obtained upon centrifugation of the supernatant at 12,000 g for 15 min is resuspended in 0.25 M sucrose and recentrifuged at 15,000 g for 20 min. The washed crude mitochondrial pellet is resuspended in 0.25 M sucrose (2 ml/g original tissue) and used for the isolation of synaptosomes (29). A gradient tube is prepared by successive layering of 5 ml (when Spinco Rotor SW 25.1 is used, and 8 ml when SW 25.2 is used) each of 20%, 16%, 12%, 8%, and 2% Ficoll solutions containing 0.25 M sucrose into a 1 in. x 3 in. cellulose tube designed for the Spinco Rotor SW 25.1. The tube is allowed to stand for 1 hr at room temperature and for 30 min at 4°. Five milliliters of the washed crude mitochondrial fraction are carefully layered on the top of the Ficoll gradient of each of the three tubes, and the contents are centrifuged at 90,000 g for 1 hr in a Spinco Model L₂-65 B at 0°. At the end of the run one obtains six distinct layers, and the synaptosomal fraction (layers B and C, Fig. 2, which peaks around 8% Ficoll, about 1.08 g/ml) is removed by means of a density gradient withdrawing device (Fig. 3). Day et al. (30) working with brain homogenates used isoosmotic Ficoll-sucrose density gradients in the B-XV rotor of an Anderson NIH-AEC zonal centrifuge and showed the synaptosomal fraction to peak at densities of 1.072 and 1.152 g/ml. By substituting isoosmotic Ficoll-sucrose for hyperosmotic sucrose, they were able to stabilize the zonal centrifuge absorbancy profiles of adult rat brain homogenates. They observed that the reason for the instability in ordinary sucrose gradients is the interaction of myelin with other brain structures in hyperosmotic sucrose. The synaptosomal fraction is carefully diluted with an equal volume of 0.25 M sucrose and centrifuged at 120,000 g for 15 min. The resulting pellets are resuspended in 0.25 M sucrose or 0.01 M Tris buffer, pH 7.4 depending on the objective of the experiment (see below).

E. Morphological and Biochemical Characterization of the Synaptosomal Fraction

The synaptosomal fraction can be monitored for purity and homogeneity as follows:

(a) Gross examination: In contrast to the myelin layer which has an opalescent appearance, the synaptosomes have a distinct yellow tinge.

(b) Electron microscopy: Electron microscopy is an important tool in ascertaining the purity of subcellular fractions. Briefly, the synapto-somal pellets are fixed overnight in 3% glutaraldehyde buffered with 0.1 M sodium cacodylate, pH 7.4, then rinsed for 3 hr in 5% sucrose buffered with 0.1 M sodium cacodylate. The pellets are post-fixed in Palade's 1% osmium tetroxide for 1 hr, then dehydrated in ethyl alcohol, propylene oxide, and finally embedded in Epon 812. Sections are cut on a Porter-Blum Ultramicrotome MT1 and stained in uranyl acetate and lead citrate. The specimens are examined and photographed with Siemens Elmiskip I A. Electron micrographs for the synaptosomal fraction should show a fairly homogeneous and intact preparation, and should consist mainly of synapto-somal particles, with synaptic vesicles, intraneuronal mitochondria and entrapped cytoplasm enclosed within an intact synaptosomal membrane. Typical electron micrographs of synaptosomes and mitochondria prepared from crude mitochondrial fraction by means of density gradient in a Ficoll-sucrose medium are shown in Figs. 4 and 5 respectively.

(c) Enzymatic markers: Succinic dehydrogenase (29, 31) and 2,4-dinitrophenol-stimulated ATPase (9) are used as markers for nitochrondria; lactic dehydrogenase and choline acetylase for the entrapped cytoplasm (29, 31, 32); Na^+-K^+-ATPase and acetylcholinesterase for the synaptosomal membrane (33, 34); NADPH-cytochrome C reductase (35) and NAD-nuc-leosidase (35a) for microsomes. The activity of some of these enzymes in the various subcellular fractions are shown in Table 1 (pp. 158-159). The main contaminant in the synaptosomal preparation could be the microsomal vesicles which are thought to originate from the endoplasmic reticulum.

(d) Pharmacological: Acetylcholine is bioassayed in the subcellular fractions by using the frog's abdominus muscle (33, 36) or the dorsal mus-cle of the leech (32, see also Ref. 4a). Another technique which we have used in our laboratory is to inject ^{3}H-choline intravenously into rats, then sacrifice the animals and isolate the brain subcellular fractions. The acetylcholine is extracted by acidification of the synaptosomal preparation, to which 10^{-4} M physostigmine was added, to pH 4.0 with dilute HCl, boiling for 10 min at 100° and assaying for labelel acetylcholine in the supernatant.

III. SYNAPTOSOMAL Na^+-K^+-ATPase

A. Kinase and Phosphatase Activities of Na^+-K^+-ATPase

The synaptosomal membrane represents an excellent example of a neuronal membrane of known morphological origin and, as expected from its functional activity, it is enriched with the ouabain-sensitive Na^+-K^+-

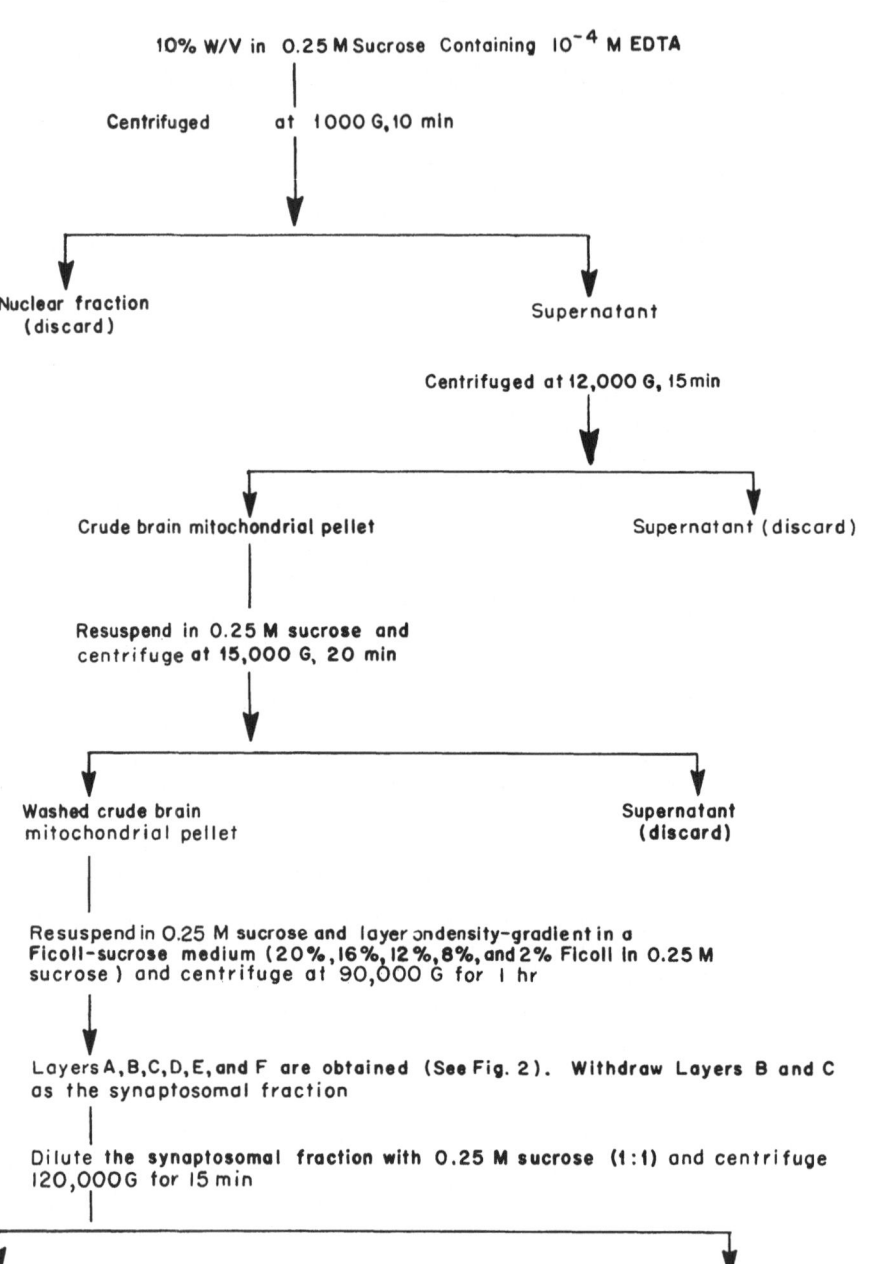

FIG. 1. Procedure for isolation of rat brain synaptosomes.

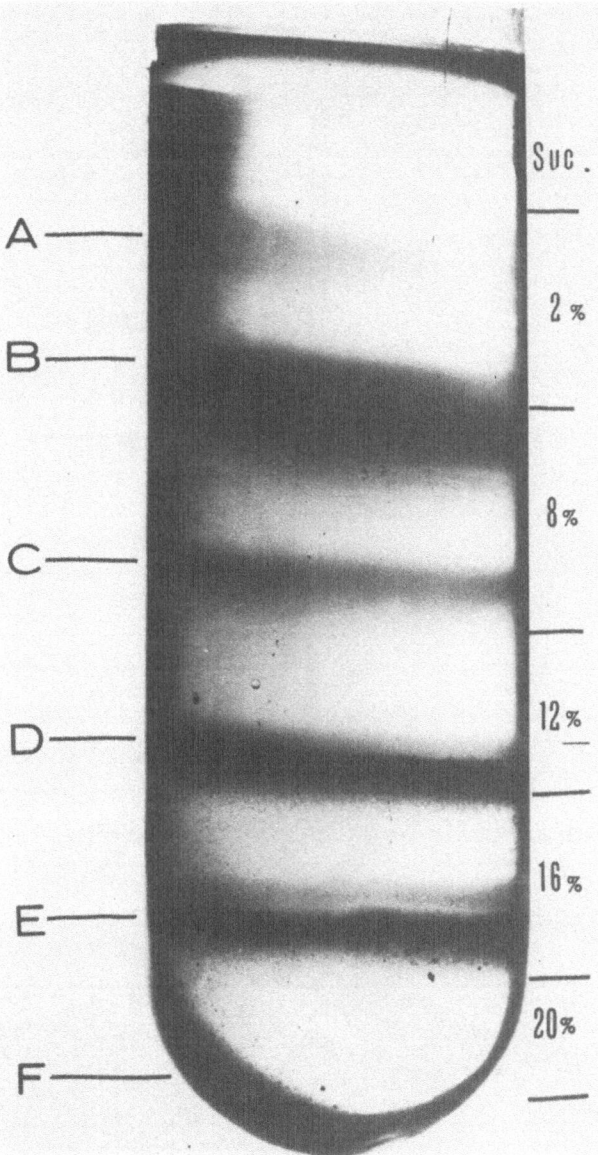

FIG. 2. Photograph of a tube showing six fractions, A, B, C, D, E, and F, after centrifuging the crude mitochondrial fraction from rat brain into a Ficoll gradient (2-20%) at 90,000 g for 1 hr.

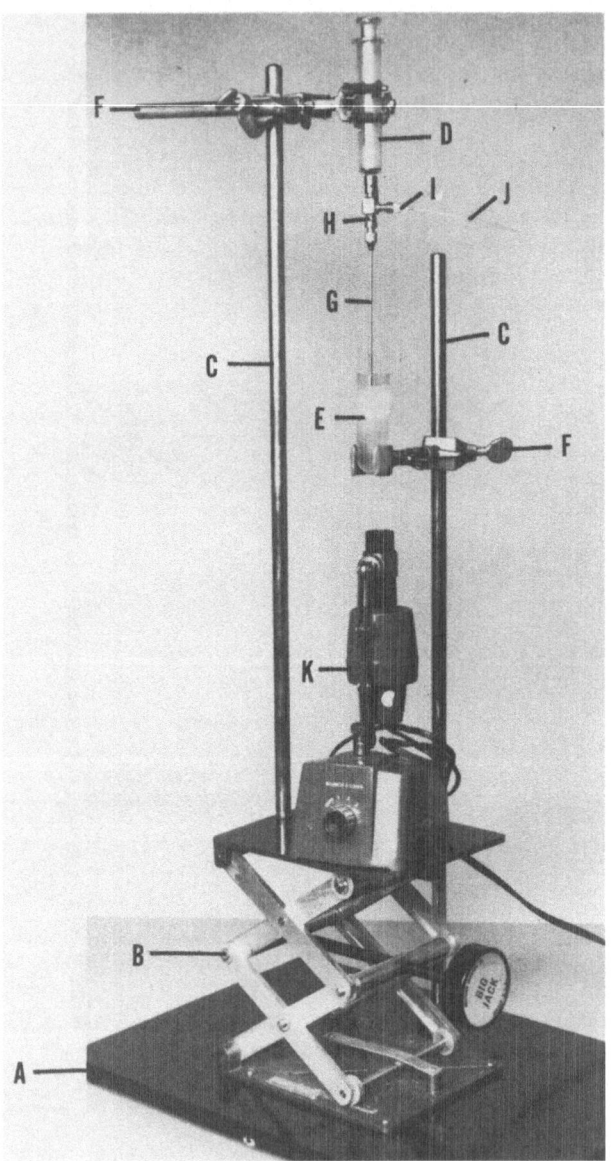

FIG. 3. Density-gradient withdrawing device. The device consists of
(A) an 45-cm long x 30-cm wide steel plate positioned on it a 60 cm support
rod (C) and a heavy-duty laboratory support jack (B), with another support
rod attached to the loading platform; a 10-ml syring with luerlok tip (D)
and the centrifuge tube (E) are held by two utility clamps (F) attached to

ATPase ($\underline{8}$–$\underline{16}$). The latter is part of the enzyme system involved in the stoichiometric transport of Na^+ outward and K^+ inward across cell membranes. It can be said in general that organs which transport Na^+ and K^+ to energize electrical activity show a rather high amount of Na^+-K^+-ATPase activity ($\underline{37}$, $\underline{38}$). Studies from several laboratories ($\underline{10}$, $\underline{13}$, $\underline{15}$, $\underline{39}$–$\underline{45}$) on the molecular basis underlying the active transport of Na^+ and K^+ and the role this enzyme system plays in this mechanism can be summarized as follows:

Na^+-dependent kinase activity   (1)

$$ATPase + Ad-Rib-P-P-*P + Na^+_{in} \xrightarrow{Mg^{2+}} Na^+_{in} ATPase-*P + Ad-Rib-P-P$$

K^+-dependent, ouabain-sensitive phosphatase activity   (2)

$$Na^+_{in} ATPase-*P + K^+_{out} \xrightarrow{Mg^{2+}} *Pi + Na^+_{out} + K^+_{in} + ATPase$$

Where Ad-Rib-P-P-*P = radioactive ATP; "in" denotes inside the synaptosome, and "out" denotes outside the synaptosome in the reaction medium.

Reactions (1) and (2) constitute the Na^+-K^+-activity observed in tissues which are known to transport Na^+ and K^+. The Na^+-dependent formation of the phosphorprotein [reaction (1)] and the decrease in its amount in-

---

the support rods on the loading platform and steel plate respectively; a syringe needle(G) luerlok, slip-on type of stainless steel, long bevel and frontal opening, 10-cm long in which the bottom end was sealed with a ball filling and two holes were drilled just above the sealed-bottom end is connected to the syringe through a three-way stopcock (H); part of a stainless disposable syringe needle (I) connects through its plastic hub a polyethylene tubing (J), 1 mm diameter, to the stopcock, and a micro lamp (K), Bausch and Lomb (Nicholas Illumination) is attached directly to the loading platform of the jack. To withdraw any layer of the submitochondrial fractions without disturbing the other layers simply adjust the microlamp (Fig. 3-K) such that the layers (Fig. 3-E) can be seen clearly. Then switch the lever of the stopcock (Fig. 3-H) to side position and lower the jack (Fig. 3-B) such that the needle openings is just below the layer to be withdrawn. Finally begin withdrawing the material with the syringe (Fig. 3-D) while simultaneously raising the jack. After withdrawing the desired layer from the tube, switch the stopcock lever to the opposite side position, place a collecting tube under the polyethylene tubing (Fig. 3-J), and empty the syringe. This procedure is repeated for each layer when all the layers are desired.

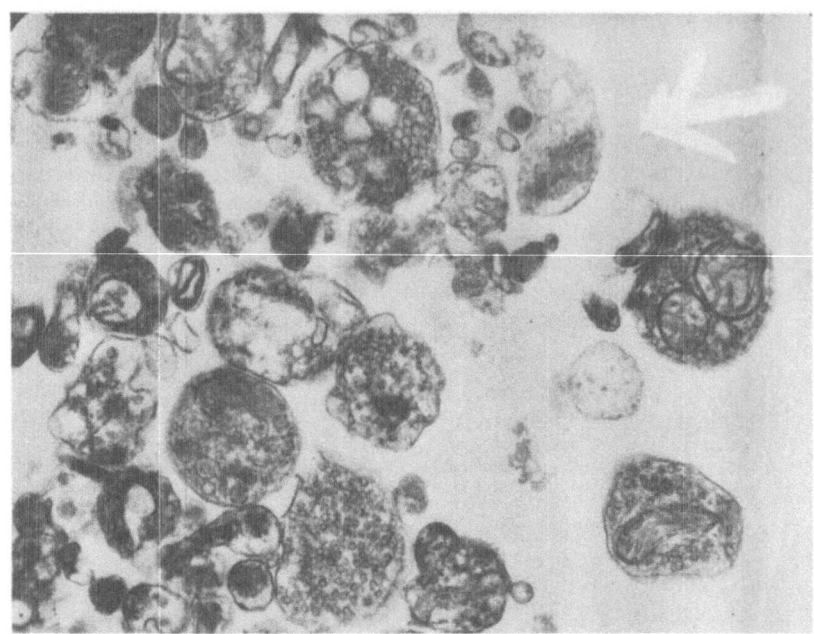

FIG. 4. Electron micrograph of the synaptosomal fraction (Layers B and C of Fig. 2). Magnification x 42,000

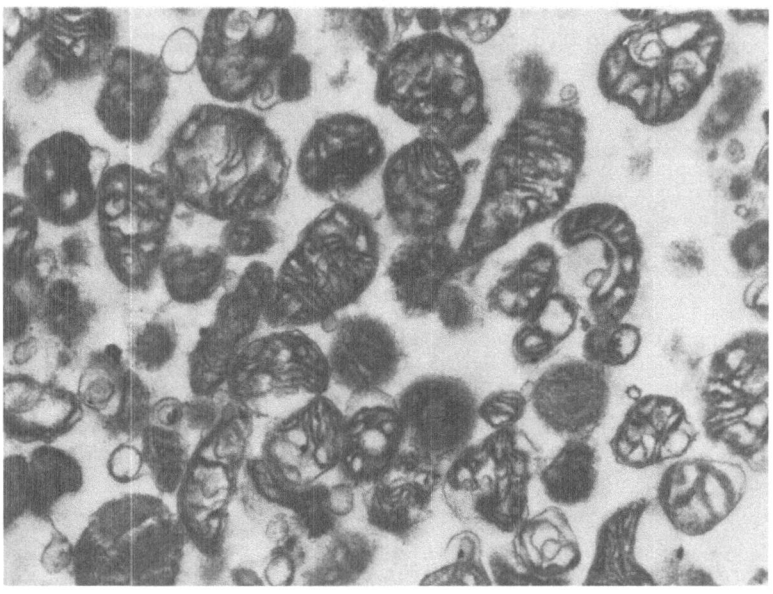

FIG. 5. Electron micrograph of the mitochondrial fraction (Layer F of Fig. 2). Magnification x 35,000.

duced by K^+ [reaction (2)] have been shown by Hokin et al. (46) to be inhibited by ouabain. Furthermore, these authors found two cathodally
moving components in the product obtained after peptic digestion and suggested that the phosphate be attached to a carboxyl group, probably in the
sidechain of an aspartic acid or glutamic acid residue. The nature of the
bond between phosphate and ATPase is still not settled at the present time.
The presence of K^+-phosphatase has also been demonstrated in the
synaptosomes (10, 13, 15, 45). Since the electrical excitability of nervous
tissue depends on the maintenance of Na^+ and K^+ concentration gradients
across the cell membrane (47), studies on the involvement of this enzyme
system in the transport of these cations in synaptosomes should shed more
light on the permeability properties of the synaptosomal membrane and
the mechanism of neurotransmission. The involvement of Na^+-K^+-ATPase
in the neurotransmission process could be depicted as shown in Fig. 6.
The Na^+-K^+-ATPase is linked to the recovery process. This process
maintains the low intracellular sodium concentration. Sodium ions are
pumped out of the cell against a high concentration and electrical gradient,
and the presence of extracellular K^+ is required for ejection of Na^+ from
the cell. Thus when 3 Na^+ move out, 2 K^+ move in and one ATP is hydrolyzed (51).

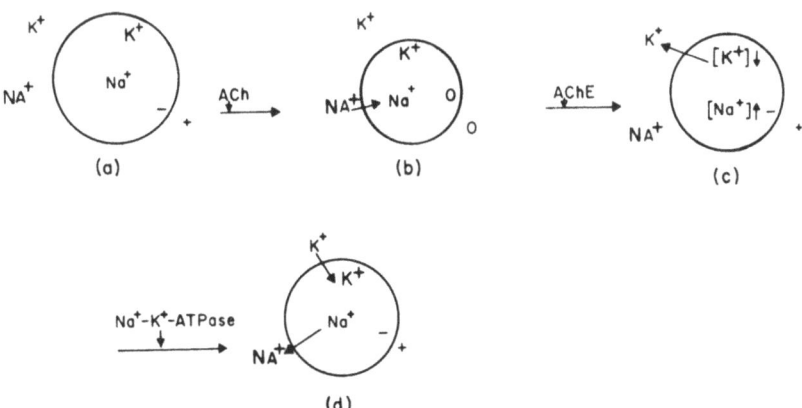

FIG. 6. Possible involvement of Na^+-K^+-ATPase in synaptic neurotransmission. (a) Resting state, membrane is polarized. (b) Upon
stimulation by acetylcholine the membrane is depolarized. (c) Following
stimulation, acetylcholine is inactivated by acetylcholinesterase and the
membrane is repolarized. (d) During recovery the membrane is polarized.
The Na^+-K^+-ATPase is involved in maintaining the concentration gradient
of Na^+ and K^+, the electrical gradient, and the cell volume at steady state
conditions.

TABLE 1

The Distribution of Some Enzymatic Markers in the Subcellular Fractions of 19-Day-Old Rat Cerebra (35b)

Subcellular fraction	Specific activity of various enzymatic markers (Δ O.D./hr/mg of protein)					
	Succinic dehydrogenase[a]	NADPH-cytochrome reductase[b]	Na^+-K^+-ATPase[c]	NAD nucleosidase[d]	Acetylcholine-esterase[e]	Latic dehydrogenase[f]
Mitochondria	4.70 (5.15)[g]	1.10 (0.63)	2.60 (0.84)	0.7 (0)	0.55 (0.18)	40.2 (60)
Synaptosomes	0.61	2.21	11.20	0.08	0.60	53.40
15-K Fraction[h]	0.16	3.74	6.50	0.16	0.95	47.40
Microsomes	0.15	6.92	8.92	0.36	1.87	41.10

All enzymes in the various subfractions were assayed in dilute solutions so that the initial rate was linear with respect to enzyme concentration.

aThe activity which reduces $K_3[Fe (CN)_6]$ at 20°C is defined as the decrease in O.D. at 400 nm/mg protein in the subfraction/hr.

bThe activity which reduces cytochrome c at 20°C is defined as the increase in O.C. at 550 nm/mg protein in the subfraction/hr.

cThe activities which liberate Pi from ATP in the presence and absence of $Na^+ + K^+$ at 37° is defined as the difference in increase in O.D. of these activities at 660 nm/mg protein of the subfraction/hr.

dThe activity which hydrolyzes NAD at 37°C is defined as the decrease in O.D. at 327 nm/mg protein of the subfraction/hr.

eThe activity which hydrolyzes acetylcholine at 37°C is defined as the decrease in O.D. of the ferric-acethy-droxamic acid complex at 540 nm/mg protein of the subfraction/hr.

f The activity which oxidizes NADH at 20°C is defined as the decrease in O.D. at 340 nm/mg protein of the subfraction/hr.

gThe values in brackets for mitochondria were obtained by passing the purified mitochondria obtained from the first density gradient on a second similar gradient. All values reported here are means of two different experiments and the average deviation ranged from 1-5%.

hObtained by centrifuging the post-mitochondrial supernatant at 15,000 g.

B. Assay Methods for Na^+-K^+-ATPase and Associated Activities
in Synaptosomes

Principle: $ATP \xrightarrow[Mg^{2+}, Na^+, K^+]{ATPase} + ADP + Pi$ (3)

The Na^+-K^+-ATPase activity can be measured as the difference in the
rate of release of inorganic phosphate from ATP in the presence of
$[Na^+ + K^+ + Mg^{2+}]$ and $[Mg^{2+} + 10^{-4}$ M ouabain].

1. Reagents

    0.02 M ATP (as Tris salt) [available from Sigma, St. Louis]
    0.06 M $MgCl_2$
    0.20 M KCl
    1.5 M NaCl
    0.05 M Tris buffer, pH 7.4
    3×10^{-4} M ouabain
    synaptosomes
    10% trichloroacetic acid (TCA)

2. Procedure

    The synaptosomes are first lysed by homogenizing in deionized water
and the precipitate obtained upon centrifugation at 120,000 g for 30 min is
suspended in 0.01 M Tris buffer, pH 7.4 to give 1.5 mg protein/ml.
Proteins are determined by the method of Lowry et al. (48) and crystalline
bovine plasma albumin is used as standard. Into three 15 x 95 mm thick-
walled test tubes in ice is added 0.4 ml of a solution of 4 mM ATP (as
Tris salt), 5 mM $MgCl_2$, and 50 mM Tris buffer, pH 7.4. To each tube
is added 0.2 ml of a well-suspended synaptosomal preparation (the lysed
synaptosomes are homogenized in 0.01 M Tris buffer just prior to addition),
approximately 0.25-0.4 mg protein, which will release 0.30-0.80 μmole
of inorganic phosphate upon incubation for 5 min. To the first tube is
added 0.1 ml each of 0.2 M KCl and 1.5 M NaCl. To the second tube is
added 0.1 ml of 5×10^{-3} M ouabain. The final volume is brought up to 1
ml by adding deionized water. The first two tubes are incubated for 5 min
at 37°. One milliliter of iced 10% TCA is added to the third tube and is
kept in ice. At the end of the run the first two tubes are also placed in ice
and one ml of 10% TCA is added to each. All assays should be carried out
in triplicate. The tubes are then centrifuged at 800 g for 15 min and the
liberated orthophosphate is determined on 1-ml aliquots of the supernatant
by either the method of Gomori (49) or Fiske and Subbarow (50). The
third tube serves as a control. Using the Gomori method (49), to the 1 ml

supernatant is added 2.8 ml water, 0.2 ml 10 N H_2SO_4, 0.5 ml 2.5% ammonium molybdate, and 0.5 ml of a solution of Elon (monomethyl-p-aminophenol sulfate, Kodak, 1 g/100 ml) and $NaHSO_3$ (3 g/100 ml). Let stand for 10 min then read at 660 nm. Reagent blanks and inorganic phosphate standard are included in each experiment and serve to convert optical density into micromoles of inorganic phosphate released per milligram of protein per hour at 37°, after subtracting the control. The control permits correction for inorganic phosphate in the synaptosome and for the non-enzymic hydrolysis of ATP. Specific activity of ATPase is defined as μmoles inorgainc phosphate liberated/mg protein/hr. The activity in the first tube ($Na^+ + K^+ + Mg^{2+}$) less that in the second tube (Mg^{2+} + ouabain) constitutes the Na^+-K^+-ATPase activity. One unit of activity splits 1 μmole of ATP per minute at 37°. Activity ratios can be calculated by dividing the Na^+-K^+-stimulated-Mg^{2+}-dependent ATPase by the Mg^{2+}-dependent ATPase in the presence of ouabain. Thus if the specific activity of synaptosomal Na^+-K^+-Mg^{2+}-ATPase is 17 and that of Mg^{2+}-ATPase is 8.5, the activity ratio is 2 and the (Na^+-K^+) increment is 8.5 (14).

C. Properties

The properties of Na^+-K^+-ATPase from various tissues have been determined and found to be quite similar (51). We have compared the properties of Mg^{2+}-ATPase and Na^+-K^+-ATPase in synaptosomes and microsomes and found striking similarities in regard to K_m, V_{max}, pH optima, and temperature optima (52). Only ATP and dATP served as active substrates for Na^+-K^+-ATPase (14).

Ebel et al. (53) showed the activities of Mg^{2+}-ATPase and Na^+-K^+-ATPase to increase in the synaptosomes after adrenalectomy and to decrease in the microsomes. Physiological doses of aldosterone diminished the increased Na^+-K^+-ATPase activity, without bringing it back to control levels. Treatment with methylthiouracil or thyroxine exerted no influence on ATPase activity in synaptosomes and microsomes. Atkinson et al. (54) purified Na^+-K^+-ATPase from synaptosomes and microsomes and showed similarities in their properties (see below). Stability: The synaptosomal Na^+-K^+-ATPase is stable for more than 3 months when kept at -20°.

D. Activators and Inhibitors

Neither Ca^{2+} nor Cd^{2+} can substitute for Mg^{2+} in the synaptosomal ATPase system, Mn^{2+} substitutes partially; only in the presence of Mg^{2+} is there activation by Na^+ and K^+ (14). The enzyme is inhibited by ouabain, fluoride, and oligomycin (14) and inhibited up to 80% by p-chlormercuribenzoate (9).

E. Purification Procedure

Recently Atkinson et al. (54) modified techniques previously described (55, 56) for purification of Na^+-K^+-ATPase of microsomes and synaptosomes from pig brain. Water-soluble "transport ATPase" was obtained by extraction of synaptosomal membranes with "Lubrol W" after treatment with deoxycholate and NaI, then purified by means of chromatography on "Sephadex G-200" and electrophoresis. The purified preparation had a molecular weight of 280,000 and isoelectric point pH of 5.0 ± 0.1, one protein band in polyacrylamide electrophoresis at pH 7.0, and specific activities (μmole Pi released/mg protein/hr) for Na^+-K^+-ATPase and Mg^{2+}-ATPase 513 ± 20 and 26.8 ± 3.8, respectively. The authors (54) suggested that the 280,000 molecular weight transport ATPase probably comprises twelve subunits of molecular weight 25,000 arranged as three tetramers of molecular weight 98,000. These findings could explain previous reports from several laboratories on the complex kinetics of the transport ATPase system (7, 57-59). Addition of phospholipids such as phosphatidylcholine and phosphatidyl serine exert a stimulatory effect on the solubilized Na^+-K^+-ATPase system (54, 60, 61).

F. Associated Activities

In addition to the Na^+-K^+-ATPase the synaptosomal membrane contains also a $(K^+ + Mg^{2+})$-dependent and ouabain-sensitive p-nitrophenyl phosphatase (K^+-Mg^{2+}-NPPase); they appear to be two distinct enzymes with different structural associations (10). Tanaka and Mitsumata (44) isolated three types of NPPases from bovine cerbral cortical membranes (acid, Mg^{2+}-activated, and K^+, Mg^{2+}-activated), and of these only K^+-Mg^{2+}-NPPase was stimulated about twofold by phospholipids. It is not clear at the present time whether the K^+-Mg^{2+}-activated NPPase is involved in the final step of the reaction sequence catalyzed by Na^+-K^+-ATPase. It is possible that the two activities are due to the existence of two different active sites.

The synaptic vesicles possess a characteristic type of ATPase which is stimulated by either Mg^{2+} or Ca^{2+} but is not effected by Na^+ and K^+, dinitrophenol, or sodium azide (62). Synaptosomes possess also adenylate kinase which could be easily extracted with 0.01 M Tris buffer, pH 7.4 (14).

IV. TECHNIQUES USED IN STUDIES ON ION TRANSPORT IN SYNAPTOSOMES

In contrast to the huge amount of information on ion transport in mitochondria, nuclei, erythrocytes, and tissue slices, our information on ion

movements in subcellular fractions derived from nervous tissue is still
scant. This is partly due to the unusual heterogeneity of brain tissue which
made it difficult in the past to prepare purified nuclei and mitochondria by
the conventional methods of homogenization and differential centrifugation.
Thus, earlier, brain mitochondria produced by conventional procedures
were found to be capable of carrying on glycolysis in addition to oxidative
phosphorylation (63). Following the development of techniques for the purification of mitochondria from the crude mitochondrial fraction by means
of density gradient centrifugation (34, 64), it became evident that the glycolytic activity is due to the presence of synaptosomes in these preparations
(31, 65, 66). Later work on the metabolism of synaptosomes showed that
these particles are capable of several metabolic activities including glycolysis and oxidative phosphorylation (29, 65, 67-70); incorporation of ^{32}Pi
into phospholipids, phosphoproteins, and nucleotides both in vivo (71) and
in vitro (72, 73); incorporation of choline into phosphatidylcholine (20, 74),
acetylcholine, phosphorylcholine, and betaine (4, 5, 19); synthesis of their
own proteins (75, 76) and synthesis of catecholamines and γ-aminobutyric
acid (1, 2, 4, 5). Furthermore, Bradford (77)[†] showed that when synaptosomes are incubated in metabolic media appropriate for whole cells, they
responded to electrical pulses of alternating polarity with accelerated
respiration and glycolysis and release to the medium a proportion of their
amino acids, especially glutamate, aspartate, and γ-aminobutyric acid.
These responses were interpreted as occurring as a result of depolarization of a trans-membrane potential generated by the snyaptosomes. Elec-
tronmicroscopic studies from several laboratories show the contents of
the synaptosome to be enclosed within a unit layer membrane (1, 4, 78).
All of these studies are consistent with the view that the synaptosome is
similar in many respects to a nonnucleated cell. Since synaptosomes
represent a specialized part of the nerve cell axon a number of investigators have studies the properties of ion movements in this preparation
during the past few years.

The following sections present the main techniques used in studies on
ion transport in synaptosomes. They are divided according to the method
employed to separate the synaptosome from its incubation mixture.

A. Gel Filtration through Sephadex Columns

Gel filtration through Sephadex G-50 (bead form) was employed by
Marchbanks (17, 18) for the rapid separation of synaptosomes from the incubation medium. The medium for choline uptake consisted of isotonic

---

[†] See also H. F. Bradford, Methods of Neurochemistry (R. Fried, ed.),
Vol. 3, Dekker, New York, 1972.

salt solution which was found to give optimum respiration rates. Sucrose
was added when required to maintain the osmolarity of the medium at 320-
420 mOs. Optimum conditions for movement of ions into synaptosomes
have not yet been worked out. Various investigators have used different
incubation media; only those used in the author's laboratory are described
in detail, as comparative studies on the merit of each method have not
been carried out.

In general, the uptake of the low molecular weight substances into
synaptosomes is determined by using small columns (11 cm high x 0.8 cm
diameter) containing 0.5 g of Sephadex G-50, previously equilibrated with
sucrose isoosmotic with the incubation mixture. At the end of incubation,
a portion of the suspension is placed on the column, and the synaptosomes
are eluted with sucrose. The eluate contains 80% of the protein and less
than 1% of the unbound small molecules applied to the column, and is taken
to represent the void volume. For control, another portion of the suspension is passed through a similar column, pre-equilibrated with water instead of sucrose, which is strongly hypoosmotic to synaptosomes. The
disrupted synaptosomes are then eluted with water and their small molecular weight substances are retained in the column. The difference between
the amount of small molecular weight substance present in the void-volume
effluent from the isoosmotically eluted column and that in the effluent from
the hypoosmotic column is termed the osmotically sensitive content of the
synaptosome preparation (17). From these studies it was concluded that
the permeability of the synaptosomal limiting membrane to potassium and
sodium ions is similar to that of other biological membranes, and suggested
that the limiting membrane is substantially intact (79). Also, their findings
on the similarities of the choline-uptake process in synaptosomes to that
in erythrocytes and brain cortex slices indicate that the synaptosomes
limiting membrane is functionally competent in this respect.

B. Filtration through Millipore Filters

This technique was used in studies on the transport of sodium (21, 80),
potassium (22), calcium (28), and tryptophan (24) across the synaptosomal
membrane. The uptake of ^{22}Na into synaptosomes is discussed in detail
in this review.

1. Principle

Filtration analysis takes advantage of the fact that a Millipore filter
can concentrate in an area small enough for convenient examination all
particulate matter larger than the filter pore size, from a large volume
of liquid. The synaptosome ranges between 0.5-1.0 μ in diameter, and if

an incubation mixture is passed through a Millipore filter with 0.45–0.8 μ in diameter, the synaptosomes with their entrapped contents are retained on the filter and the latter is analyzed.

2. Reagents and Equipment

Stock solution containing (mM): NaCl, 280; Na H_2PO_4, 20; $CaCl_2$, 2.4; $MgSO_4$, 2.0; glucose, 20.0; and approximately 20 mM NaOH to adjust the pH of the solution to 7.4.

[22]Na as [22]NaCl, purchased from Amersham/Searle Corporation.
Freshly prepared synaptosomes.
Millipore filters, 25 mm in diameter and 0.45 μ pore size.
Pyrex microanalysis filter holder, with fitted glass base, Pyrex funnel and base, anodized aluminum clamp, and 125-ml filtering flask with one-hole stopper to fit (all can be purchased from Millipore Filter Corp., Bedford, Massachusetts).
Tri-carb liquid scintillation counter.
Scintillation liquid, toluene-ethyl alcohol (6:4, v/v) as the solvent, and PPO and POPOP as scintillators.

3. Sodium-uptake Experiments

The standard uptake mixture is made at 0° by adding the various components in the following order: (1) 1.0 ml of a suspension of synaptosomes from 2–2.5 g of brain in 0.25 M sucrose, (2) 1.0 ml of the stock salt solution, and (3) 1.0 ml of a [22]Na solution containing 2 μCi of the radioisotope. Immediately after mixing, 0.1 ml of the incubation mixture is taken as the zero-time sample and the test tube containing the uptake mixture is dipped with gentle shaking into a 37° water bath to attain temperature equilibration. Samples (0.1 ml) are removed from the reaction mixture at various time intervals, suspended in 5 ml of precooled washing solution (the stock salt solution diluted with an equal volume of deionized water), rapidly mixed by shaking gently, and immediately filtered under vacuum through a Millipore filter 25 mm in diameter and 0.45 μpore in size. The Millipore filter is mounted on a fritted disk with a 25-ml capacity funnel fitted over it. An additional 10 ml of the washing solution is used to wash the synaptosomes on the Millipore filter. Suction is continued until the filter is dry. The filter paper is dried under infrared light and counted in a Tri-carb liquid scintillation counter with toluene-ethyl alcohol (6:4, v/v) as the solvent and PPO and POPOP as scintillators. A filter paper, previously wetted by passing 10 ml of the above filtrate (total washings used in suspension and washings of the particles) through it, serves as the blank. Using this method, one can study the effect of varying the incubation conditions as well as the effects of ions, metabolic inhibitors, and drugs on the uptake of cations into the synaptosomes (21, 22, 28).

C. Filtration through Amberlite IRC-50 (Na$^+$) Cation Exchange Resin

This technique was used in studies on the transport of choline into synaptosomes (19, 20, 74).

1. Principle

The Amberlite IRC-50 (Na$^+$) cation exchange resin (25 to 50 mesh, can be purchased from Mallinkrodt Chemical Co., St. Louis) is coarse and thus does not impede the flow of synaptosomes through the column. The choline in solution exchanges with the sodium on the column while that inside the synaptosomes and unavailable to the resin is recovered in the effluent. The transport is determined by measuring the radioactivity in the supernatant of the lysed synaptosomes.

2. Reagents and Equipment

^{3}H- or ^{14}C-Methylcholine (Purchased from New England Nuclear)
    Amberlite IRC-50 treated with 1 N NaOH, then washed with de-
    ionized water until pH reaches 7.5.
Isoosmotic incubation medium (see below under incubation conditions).
High voltage electrophoresis.
Varsol. It is a mixture of hydrocarbons, purchased from Humble Oil
    Co.
1.5 M acetic acid-0.75 M formate buffer, pH 2.0.

3. Conditions of Incubation and Filtration of the Synaptosomes

The stock salt solution used for suspending the synaptosomes consisted of (mM): NaCl, 166; KCl, 8.3; Tris buffer, pH 7.4, 83; MgCl$_2$, 4.2; CaCl$_2$, 0.85; physostigmine, 1.65; mercaptoethanol, 8.3; glucose, 16.5; and sucrose, 16.5. The synaptosomes are homogenized carefully in the above salt solution such that 0.3 ml of the homogenate contains 2-2.5 mg of protein. To 0.2 ml of a solution containing 0.5 μCi (0.25 mM) of ^{14}C-methyl-labeled choline plus other additions as the experiment calls for, 0.3 ml of the synaptosome suspension is added to give a final concentration of 340-380 mOs solution. The final volume of the reaction mixture was 0.5 ml. All incubation reactions were run in triplicates. The control tubes contained 20 mM of unlabeled choline, and the uptake of radioactivity by the specific, high affinity, saturable system is prevented by the presence of the high level of unlabeled choline, whereas nonspecific leakage of the labeled molecules into the synaptosome is not affected (19). After incubation at 37° for 15 min, the tubes are chilled in ice and 1 ml of cold 0.25 M sucrose containing 20 mM unlabeled choline is added to the experimental (1 ml of 0.25 M sucrose is added to the control). The contents of the tubes are then passed over columns (10.5 cm x 0.8 cm) filled with Amberlite IRC-50 (Na^+) cation exchange resin (25-50 mesh) previously conditioned with 0.25 M sucrose. An additional 9 ml of 0.25 M sucrose is passed through the column. The eluate, which contains all of the synaptosomes added to the column, is centrifuged at 80,000 g for 30 min, the precipitate is washed with sucrose and the centrifugation repeated. To extract the contents of the synaptosomes, 1 ml of deionized water is added, containing 0.1 mM each of physostigmine, acetylcholine, phyosphorylcholine, betaine, and choline, previously adjusted to pH 4.0 with HCl, and the mixture is heated at 100° for 10 min. During these studies we found that physostigmine interfered with the protein determination (48), thus it is advisable to determine proteins prior to the addition of physostigmine. The cellulose tubes were frozen overnight. Nine milliliters of water are added to each tube, shaken vigorously, and centrifuged at 80,000 g for 30 min. The supernatant is lyophilized and dissolved in 0.2 ml of deionized water, then the radioactivity of an aliquot is assayed and the rest is run in high-voltage paper electrophoresis at pH 2.0 as described in Fig. 7. The paper is dried, exposed to iodine vapor and the spots, which correspond to phosphorylcholine, betaine, acetylcholine, and choline, are marked, cut, and counted in the liquid scintillation counter; in general, we find the distribution of radioactivity in these spots to be 12, 4, 16, and 68% respectively [Table 2 (81)]. This electrophoretic technique could easily be applied to studies on the effect of various conditions, metabolic inhibitors, and drugs on choline uptake and its subsequent metabolism in synaptosomes.

When the lysed synaptosomal precipitate (see above) are extracted with chloroform-methanol (2:1) and the lipid extract is separated into its individual components by means of two-dimensional thin-layer chromatography, nearly all the radioactivity is recovered in phosphatidylcholine. Further studies showed that ^{14}C-choline enters phosphatidylcholine by base exchange (20, 74).

D. Other Methods Used for Permeability Studies of the Synaptosome

Among the methods employed in studies on the permeability properties of the synaptosomal membrane are the sedimentation of the synaptosomes from the reaction medium by centrifugation and the spectrophotometric measurement of the rate of swelling of the particles.

1. Precipitation of the Synaptosomes from the Incubation Medium by Centrifugation

Studies on the uptake of choline and catecholamines into synaptosomes are carried out in an isoosmotic medium containing the radioactive

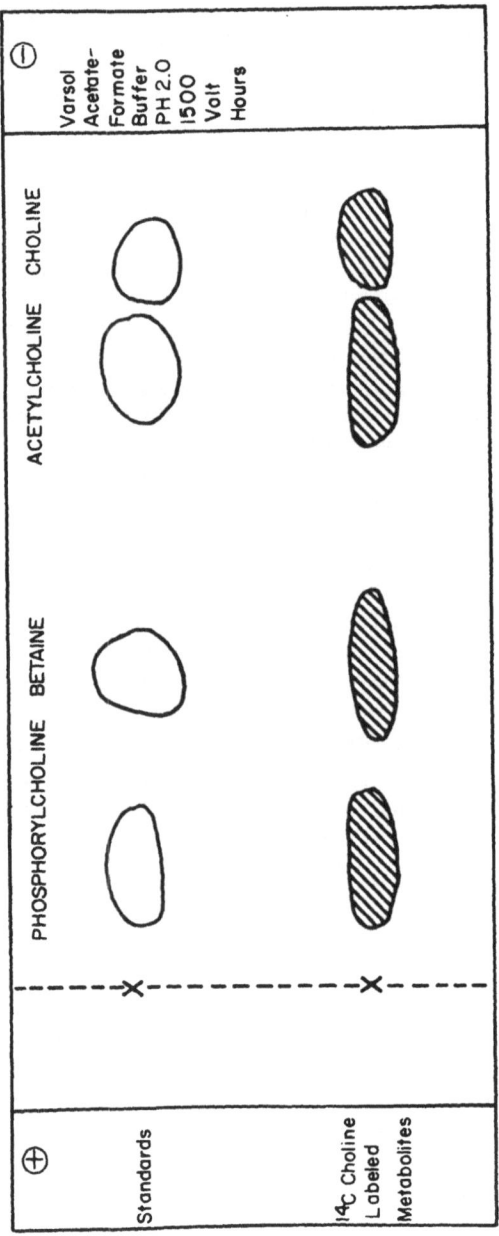

FIG. 7. Paper ionopherogram of the ^{14}C-labeled choline and its metabolites following uptake into the synaptosomes. The spots were applied along the dotted line.

TABLE 2

Uptake of ^{14}C–Choline by Rat Brain Synaptosomes

Experiment	Conditions of incubation[a]	Radioactivity in the various choline metabolites after electrophoresis of the supernatant of the lysed synaptosomes (expressed in cpm)				
		Phosphorylcholine	Betaine	Acetylcholine	Choline	
1.	^{14}C–choline (0.17 mM)	1140	382	1630	4450	
2.	^{14}C–choline (0.17 mM) + unlabeled choline (15 mM)	342	115	490	1340	
3.	^{14}C–choline (0.17 mM) + hemicholinium–3 (0.1 mM)	792	257	1140	3100	
4.	^{14}C–choline (0.17 mM) + acetylcholine (1 mM) + 0.5 mM physostigmine	923	268	1170	3470	

[a]Synaptosomes from rat brain equivalent to 2.35 mg of protein were incubated under the conditions described in the text for 15 min at 37° and assayed by passage over Amberlite IRC–50 cation exchange resin and high voltage paper electrophoresis as described in the text. Specific uptake can be calculated by subtracting the uptake in experiment 1 (Abdel–Latif and Smith, unpublished work).

neurotransmitter or its precursor and following incubation the reaction mixtures are centrifuged at high speeds and the contents are lysed and counted (18, 25-27). Potter (18) incubated synaptosomes in a balanced isotonic salt solution (82), which was saturated with 5% CO_2 in O_2, at pH 7.3 and 20° in a final volume of 100 ml. ^{14}C-Choline (1-1000 μM, specific radioactivity 40 mCi/mmole) was added to start the reaction. Two-milliliter portions were taken at various time intervals, placed in plastic centrifuge tubes, and centrifuged at 20,000 g for 30 sec in a microcentrifuge. The pellet obtained upon centrifugation was suspended in ethanolic phosphor for assay of total radioactivity or extracted with ethanol containing perchloric acid for assay of labeled choline and acetylcholine.

2. Spectrophotometric Measurement of Rate of Swelling of Synaptosomes

The light-scattering technique was used by Keen and White (23, 83) to study the permeability of rat brain synaptosomes to various ions including sodium, potassium, chloride, ammonium, calcium, magnesium, and phosphate. The rate of swelling of a particle can be measured spectrophotometrically. Dilute suspensions of synaptosomes are turbid and absorb and scatter visible light. When the synaptosomes swell the light scattering decreases and the optical density of the suspension decreases (84). The swelling can be measured spectrophotometrically by recording changes in absorbance at 520 nm. Commercial light-scattering photometers are used; fluorometers with appropriate slits could serve the same purpose. Light-scattering techniques have been used to study the permeability of erythrocytes (85) and mitochondria (86).

In brief, the rate of swelling in synaptosomes is measured as follows (83): 2.8 ml of a 0.4 M solution of the test substance is added to a 1-cm cuvet and 2 min later 0.2 ml of the synaptosome suspension is added and the extinction at 520 nm is recorded immediately. As base line for each test substance, 2.3 ml of distilled water is pipeted into a cuvet; 2 min later 0.2 ml of the synaptosome suspension is added. After 2 min, during which the synaptosomes are ruptured, 0.5 ml of 0.24 M solution of the test substance is added and the extinction is recorded at 520 nm (Fig. 8). An advantage of this technique is that it distinguishes the intrasynaptosomal ions from the extracellular or bound ions (87).

V. PERMEABILITY PROPERTIES OF THE SYNAPTOSOMAL MEMBRANE

A considerable amount of evidence has accumulated in the past few years showing the separation of the intrasynaptosomal constituents from the outside environment by a permeability barrier which has many of the general characteristics of a biological membrane. This conclusion is

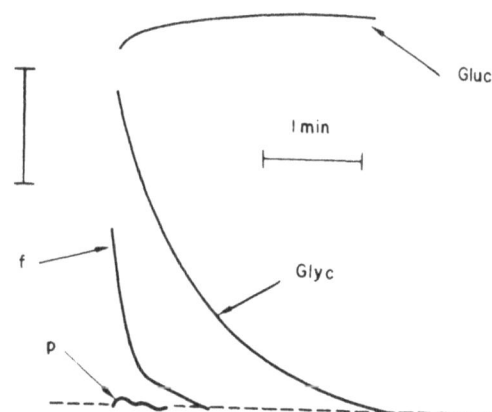

FIG. 8. Swelling of synaptosomes in 0.8 M solutions of four non-electrolytes at 27° as followed by spectrophotometry. Recordings have been superimposed. The broken line shows the absorbance of a suspension of ruptured synaptosomes. The vertical scale represent a 10% transmission change; a fall of A_{520} indicates swelling. gluc, glucose; glyc, glycerol; f, formamide; p, propylene glycol. Reprinted from Ref. 83, p. 569, by courtesy of the authors and publisher.

supported by the following general observations: (a) From the appearance of synaptosomes in thin section, it was suggested that they consist of sealed bags of cytoplasm with a continuous external plasma membrane (78). Studies from several laboratories on the morpho-biochemical properties of synaptosomes have confirmed this conclusion. (b) The synaptosomal sap contains the following components: acetylcholine, most of which is bound to the synaptic vesicles, and the rest is found in the labile form (88); choline acetylase (1, 4, 89); K^+ (90); the glycolytic enzymes (1, 3, 65–67), and enzymes of the hexose-monophosphate-shunt pathway (68). (c) Components of these particles are immediately released upon rupturing the external synaptosomal membrane by hypoosmotic shock. (d) Selective permeability to sodium and potassium ions and the presence of Na^+-K^+-ATPase, a choline transport system, and carriers for amino acids and catecholamines in the external membrane of these particles. That these preparations retain their structural and functional integrity is supported by the following specific findings:

A. Sodium and Potassium Transport

Marchbanks (4, 79) passed a suspension of synaptosomes through a Sephadex column that was eluted with an isoosmotic solution and observed about 45% of the K^+ of the preparation is recovered in the void volume of the effluent with the synaptosomes. In contrast when he eluted the column with the hypoosmotic solution, 5 mM Tris buffer, pH 7.4, K^+ ions were lost (Fig. 9). Similarly when the synaptosomal membrane was disrupted with detergents or application of supersonic oscillation, K^+ were lost. These experiments show that the osmotically labile void-volume K^+ represent K^+ entrapped within the synaptosomes and suggest that the synaptosomal membrane is substantially intact. Studies on radioactive sodium uptake into synaptosomes by Ling and Abdel-Latif (21) showed ouabain or iodacetate plus cyanide to exert an inhibitory effect on Na^+ outflux but not influx (Fig. 10). When K^+ was added to a medium containing synaptosomes loaded with Na^+ (^{22}Na), an immediate release of Na^+ from these particles was observed. These observations suggest the existence of a Na^+-K^+ exchange transport system in synaptosomes. The action of potassium was inhibited by $10^{-4}M$ ouabain at low (3.3 mM) but not at high (20 mM) K^+ concentration (Table 3). The uptake and efflux of Na^+ was temperature-dependent and only intact synaptosomes were found to transport these ions actively. Bradford (91) showed that during incubation in physiological

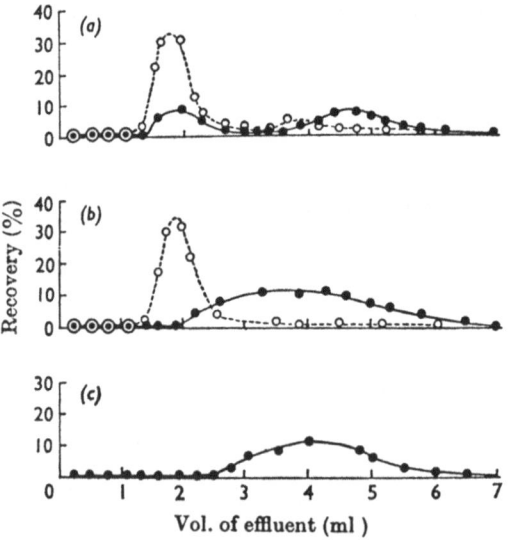

FIG. 9. Separation on Sephadex columns. ●, K^+: o, protein. (a) 0.5 ml. of synaptosome preparation eluted with 0.8 M-sucrose; (b) 0.5 ml of synaptosome preparation eluted with 5 mM-Tris-HCl buffer, pH 7.4; (c) 0.5 ml of 2.5 mM-KCl eluted with 0.8 M-sucrose. Reprinted from Ref. 79, p. 150, by courtesy of the authors and publisher.

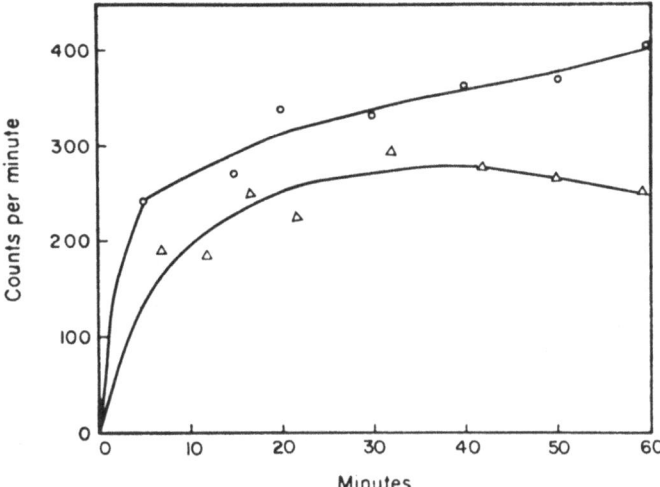

FIG. 10. Effect of KCN and iodacetate on $^{22}Na^+$-uptake by rat brain synaptosomes. The standard uptake mixture contained 2.2 mg of particle protein/ml and was incubated with 0.001 M KCN + 0.001 M-iodoacetate (upper curve) for 10 min at 37° before the addition of $^{22}Na^+$. The lower curve represents the control which was incubated withour inhibitors. Reprinted from Ref. 21, p. 725, by courtesy of the publisher.

saline synaptosomes concentrate K^+ about sevenfold as compared with a thirty-fold concentration by cortex slices. Escueta and Appel (22) studied the uptake of $^{42}K^+$ into synaptosomes and found upon the addition of 100 mM Na^+ and 10 mM K^+ an increase in K^+ concentration (measured as μmole/mg protein) from 0.092 to 0.14–0.25 after 8 min of incubation at 23°. Ouabain was an effective inhibitor only of the Na^+-dependent K^+ accumulation. These results suggest that synaptosomes support active potassium transport. Addition of ATP, ADP, α-ketoglutarate, succinate, fumarate, or glucose did not influence K^+ accumulation. Keen and White (23) showed that Gramacidin D selectively increased the permeability of synaptosomes to Na^+ and K^+. Diamond and Goldberg (28) studied the uptake of calcium ions into synaptosomes. They demonstrated the ability of microsomes and synaptosomes from rat brain to accumulate ^{45}Ca against a concentration gradient by an ATP-dependent process. Both particles accumulated Ca^{2+} to a similar extent but less actively than mitochondria.

B. Choline Transport

Neural tissue appears to depend on exogenous choline for the synthesis of acetylcholine and choline–containing phospholipids (89). An effective

TABLE 3

Effects of Ouabain and K^+ on Na^+-Uptake
by Rat Brain Synaptosomes (21)

Additions	Na^+-uptake (μmoles Na^+/mg protein)
None	0.160
Ouabain (10^{-4}M)	0.205
K^+ (20 mM)	0.110
Ouabain + K^+	0.151

choline transport system would be expected to reutilize the liberated choline at the synapse for acetylcholine synthesis. Carrier-mediated transport of choline has been reported in kidney slices (92), squid axon (93),
brain slices (94-96), erythrocytes (97, 98), and synaptosomes (17-20, 74).
Uptake of choline into synaptosomes exhibits saturation kinetics and K_T,
the concentration for the half-maximal specific transport of choline, is
8.3×10^{-5} M; choline uptake does not require energy or monovalent nor
divalent cations and is inhibited competitively by hemicholinium-3 (19).
Potter (18) and Marchbanks (17) demonstrated a Na^+-dependent, hemi-
cholinium-inhibited, carrier-mediated transport of choline across the
synaptosomal membrane. Schuberth and his co-workers (94, 95), working
on the uptake of labeled choline by mouse brain slices, interpreted their
results in terms of an energy-dependent system, sensitive to ouabain and
the concentration of extracellular sodium, in parallel with passive diffusion.
In contrast Cooke and Robinson (96) showed no effect of Na^+ on choline
uptake into rat brain slices, and Diamond and Kennedy (19) reported inhibitory effect of monovalent or divalent cations on choline uptake into
synaptosomes. These authors also reported that choline transport did
not require energy and it was inhibited by reagents which attack sulfhydryl
groups.

Recent studies in our laboratory (98a) on the influence of various ions
on choline uptake into synaptosomes showed a slight stimulation in the
presence of Mg^{2+} and a negligible effect upon the addition of Na^+, K^+,
Ca^{2+}, glucose, and pyruvate (Table 4). Routinely we have added Na^+, K^+,
Mg^{2+}, Ca^{2+}, and glucose (see conditions of incubation under Sec. IV, C
above) in order to obtain a more physiological incubation medium and to
preserve the structural integrity of the synaptosome. To show whether
energy is involved in choline transport, the effects of various inhibitors
of metabolism on choline transport and its subsequent metabolism were
investigated and the results of these experiments are shown in Fig. 11.

TABLE 4

Effect of Addition of Various Ions on Choline Uptake into Rat Brain Synaptosomes (98a)

Experiment no.	Ions added	Concentration of ions added (mM)	Total radioactivity in lysed supernatant (% of control)
1	None[a]	–	100
2	1 + Mg^{2+}	5	117
3	2 + Na$^+$	100	98
4	3 + K$^+$	5	102
5	4 + Ca^{2+}	1	97
6	5 + K$_2$HPO$_4$	5	100
7	6 + glucose	10	105
8	7 + Na–Pyruvate	5	101

[a]The incubation mixture consisted of 55 mM Tris buffer, pH 7.5; 0.025 mM ^{14}C–methyl–labeled choline, synaptosomes equivalent to 2.1 mg protein and enough sucrose to maintain the osmolarity of the medium at 320–325 moS. Other ions were added as indicated and the incubation conditions and assay for choline uptake were as described in the text. The effect of a certain ion on choline uptake was obtained by dividing the activity in the lysed supernatant in the presence of the ion by that in the preceding experiment lacking that particular ion.

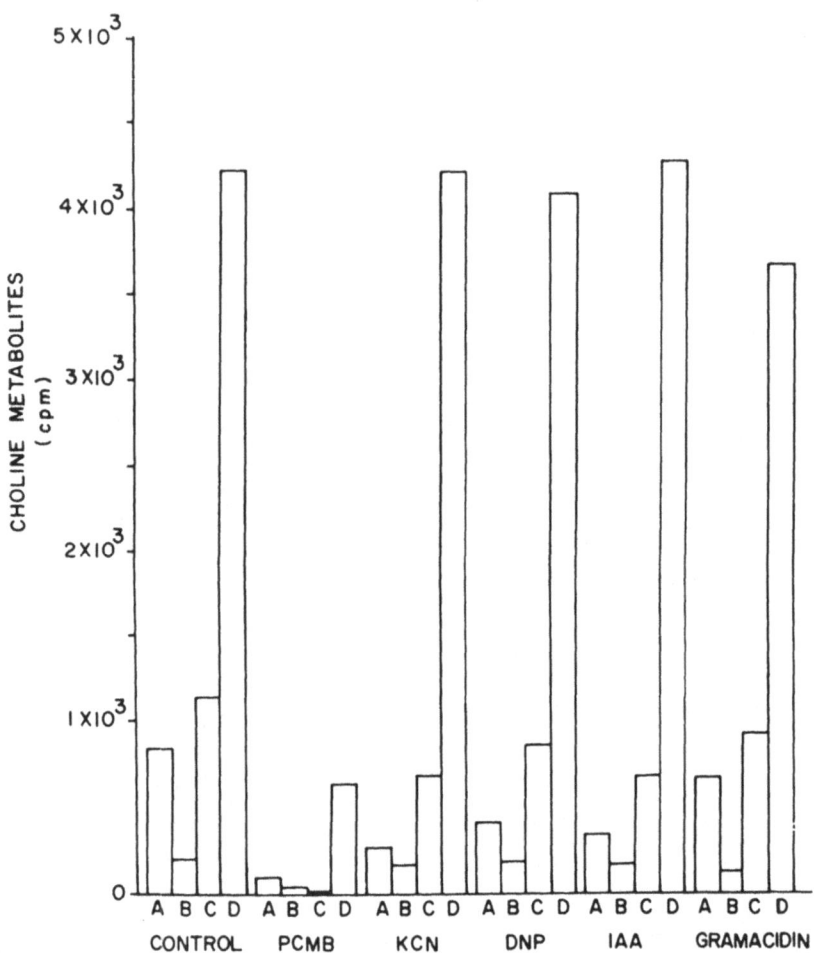

FIG. 11. Effect of metabolic inhibitors on choline uptake and metabolism in synaptosomes. Concentration of the inhibitors were as follows (mM): p-chloromercuribenzoate (PCMB), 1; Potassium cyanide (KCN), 1; 2,4-dinitrophenol (DNP), 0.2; iodoacetic acid (IAA), 1; and gramacidin, 5 μg/0.1 ml containing 17 μmoles ethanol. The letters correspond to: A, P-choline; B, betaine; C, ACh; D, choline. Conditions of incubation and analysis for choline uptake and metabolism were as described under Sec. IV, C above. Taken from Ref. 98a.

p-Chloromercuribenozoate, which attacks sulfhydryl groups, exerted up
to 85% inhibition on choline uptake as well as its incorporation into phosphatidylcholine, and virtually abolished the synthesis of acetylcholine and
to a lesser extent that of phosphorylcholine and betaine. In contrast KCN,
DNP, and IAA had little influence on choline transport and its oxidation to
betaine but, as expected, these agents inhibited considerably (23-64%)
choline conversion to acetylcholine and phosphorylcholine. The antibiotic
gramicidin, which inhibits oxidative phosphorylation and promotes the
entry of K^+, Na^+, Li^+, Rb^+, and Cs^+ into mitochondria (98b) and Na^+ and
K^+ into synaptosomes (23), exerted a slight inhibitory effect on choline
transport and metabolism (Fig. 11). Low concentrations of ethanol (30 mM)
had no influence on choline uptake and metabolism, however at higher concentrations (e.g., 600 mM) it inhibited phosphatidylcholine labeling by almost 60%, but still without much effect on choline transport. Ethanol could
compete with choline for the choline in phosphatidylcholine in the base exchange reaction. The observation that KCN, DNP, and IAA had no effect
on choline uptake (Fig. 11) is in agreement with the conclusion that the
transport of choline into synaptosomes is not energy dependent. However
these metabolic inhibitors exerted up to 40% inhibition on the conversion
of choline into acetylcholine and phosphorylcholine, which require energy
for their synthesis. Since there was little change in betaine formation and
choline accumulation in synaptosomes in the presence of these inhibitors
(Fig. 11), the marked decrease one observes in the total radioactivity of
the lysed synaptosomal supernatant could mostly be accounted for by the
inhibition of acetylcholine and phosphorylcholine synthesis. Thus the
findings of Schuberth et al. (94, 95) on the involvement of energy and Na^+
in choline transport into brain slices could be explained by assuming that
there are three pools in the nerve terminal, a choline pool, which does
not require energy, and two separate p-choline and ACh pools which require energy for their synthesis and are in equilibrium with the choline
pool. In the presence of energy and at maximal velocity for choline uptake into the nerve terminal the three pools are saturated, and the flow of
choline from the choline pool into the other two pools is at maximal rate.
When energy is interrupted by adding a metabolic inhibitor to the nerve
terminal, the flow of choline to the other two pools will stop and only the
choline pool will be saturated with the newly added choline (radioactive).
The addition of PCMB acts by inhibiting the choline transport system and
subsequently the uptake of the newly added choline by the three pools. It
must be emphasized that energy might be involved in choline transport at
much lower concentrations of choline than used in the present studies.

The choline transport system also takes up acetylcholine, although at
a lower rate, and both substrates compete for the carrier. In these experiments hydrolysis of acetylcholine by acetylcholinesterase should be

taken into consideration even in the presence of 10^{-4} M physostigmine. Recently Itokawa and Cooper (99) demonstrated the release of [^{35}S] thiamine from synaptosomes isolated from rat brain after administration of the labeled compound into the incubation medium upon the addition of acetylcholine or tetrodotoxin.

C. Norepinephrine and Tryptophan Transport

Two findings related to aminoacids uptake into whole cells (100, 101) are relevant to a discussion of norepinephrine and tryptophan transport into synaptosomes and are reviewed here. (a) In order to transport amino acids into the cell, the presence of sodium ions in the outside medium is required. The energy inherent in the downward gradient of Na^+ into the cell is believed to be the driving force for the inward transport of amino acids against a concentration gradient. However the uptake of Na^+ is linked to the active pumping of Na^+ from the cell. The energy for the latter comes from ATP. (b) Interruption of the energy required for the uptake of amino acids against a concentration gradient does not cause the cell to lose the ability to accumulate these substances. Under these conditions the amino acids are transported across the cell membrane through facilitated diffusion which does not require a Na^+ gradient and appears to be passive in nature.

The studies on norepinephrine and tryptophan uptake into synaptosomes suggest that the conditions for this transport are similar in their ionic requirements to the uptake of amino acids and sugars in other tissues (25, 26, 102). Thus, Colburn et al. (25) demonstrated the uptake of norepinephrine in synaptosomes. This was found to be Na^+- and K^+-dependent. For optimal uptake, Na^+ and K^+ were required at levels of 140 mM and 203 mM, respectively. The uptake is inhibited by reserpine (10^{-8} M) and has a maximum velocity of 0.1 $\mu g/min/g$ and a K_m of 5.6 x 10^{-7} M. Bogdanski et al. (102) showed that Na^+ is essential for the process that accumulates and stores norepinephrine and 5-hydroxytryptamine (Table 5) and that in synaptosomes the action of Na^+ is inhibited by K^+ or ouabain. White and Keen (26) also observed an increase in norepinephrine uptake into synaptosomes upon the addition of Na^+ which was depressed by 2,4-dinitrophenol and sodium cyanide. These authors concluded that an additional energy source is probably needed in addition to the energy derived from the inward-directed Na^+ concentration gradient and/or an outward-directed K^+ concentration gradient for the uptake of norepinephrine by synaptosomes.

Grahame-Smith and Parfitt (24) showed tryptophan uptake into synaptosomes to be carrier-mediated and rupture of the latter released about 97% of the amino acid unchanged. The uptake was rapid, temperature dependent, partially inhibited by cyanide and ouabain, but low external sodium

TABLE 5

Accumulation of ^{3}H-Norepinephrine and ^{14}C-Serotonin
by Synaptosomes (102)

| | Ratio of intracellular concentration to extracellular concentration | | |
| | ^{3}H-Norepinephrine | ^{14}C-Serotonin | |
mM [Na$^+$]	Control	Control	Pargyline
0	0.5	1.5	6
25	6.5	10	21
75	18	11.5	34
143	35	13	29

ion concentrations exerted little effect. Uptake of labeled tryptophan was greatly stimulated by preloading with tryptophan and was competitively inhibited by phenylalanine. These studies suggest the involvement of a saturable carrier-mediated transport system. Further evidence for a carrier-mediated transport process comes from the finding that L-tryptophan efflux from synaptosomes which were preloaded with labeled tryptophan was stimulated by cold tryptophan or any one of a number of related amino acids, a clear indication of the presence of counter-transport in these particles. The characteristics of saturability and inhibition by structural analogs have been used to distinguish carrier-mediated transport from passive diffusion. This work suggests that the limiting membrane of the synaptosomes contains carrier-proteins that can recognize and bind norepinephrine and tryptophan and facilitate their transport across the membrane. Furthermore, a supply of energy to the norepinephrine carrier, but not to that of tryptophan, is needed so that the amine can move against a concentration gradient. This energy could be applied in the form of a Na$^+$ gradient which is in turn formed at the expense of ATP through the Na$^+$-pump.

VI. CONCLUSION

From the summary of studies presented in this chapter on Na$^+$-K$^+$-ATPase and the permeability properties of the synaptosomes, it is reasonable to conclude that these particulates are enclosed within intact and functional membranes which, like other cell membranes, serve effeciently as a permeability barrier to the influx and outflux of various ions. Although it is premature at the present time to conclude that this preparation can actively transport ions against their concentration gradient, the enrichment of this preparation with Na^+-K^+-ATPase coupled with the uptake studies on Na^+, K^+, and norepinephrine suggest the presence, at least in vivo, of a Na^+-pump in this part of the nerve cell axon. Granted that conclusive experimental evidence for a Na^+-pump is still lacking, it must be emphasized that the coupling of energy to transport of ions is still not clear in slices, cells, and subcellular fractions. Thus, the techniques developed during the past ten years for the isolation of synaptosomes and studies on ion movements and transport systems for Na^+, K^+, choline, tryptophan, and norepinephrine in the membranes of these particles should serve as an impetus for exploiting this preparation for studies on the permeability properties of subcellular neuronal membranes.

ACKNOWLEDGMENT

Experimental work reported from the author's laboratory in this chapter was supported by U.S. Public Health Service Research Grant NS-07876.

REFERENCES

1. E. DeRobertis, Histophysiology of Synapses and Neurosecretion, Pergamon, New York, 1964, p. 122.

2. E. DeRobertis, Ultrastructure and cytochemistry of the synaptic region, Science, 156, 907 (1967).

3. V. P. Whittaker, The application of subcellular fractionation techniques to the study of brain function, Progr. Biophys. Mol. Biol., 15, 39 (1965).

4. V. P. Whittaker, in Structure and Function of Nervous Tissue (G. H. Bourne, ed.), Vol. 3, Academic, New York, 1969, p. 1.

4a. V. P. Whittaker and L. A. Barker, "The Subcellular Fractionation of Brain Tissue with Special Reference to the Preparation of Synaptosomes and Their Component Organelles," Methods of Neurochemistry (R. Fried, ed.), Vol. 2, Dekker, New York, 1972, Chap. 1.

5. R. M. Marchbanks and V. P. Whittaker, in Biological Basis of Medicine (E. E. Bittar and N. Bittar, eds.), Vol. 5, Academic, New York, 1969, p. 39.

6. V. P. Whittaker, Membrane phenomena at the synapse, Neurosciences Res. Progr. Bull., 9, No. 3, 387 (1971).

7. J. C. Skou, The influence of some cations on an ATPase from peripheral nerves, Biochim. Biophys. Acta, 23, 394 (1957).

8. M. Kurokawa, J. Sakamoto, and M. Kato, Distribution of Na^+-K^+-ATPase activity in isolated nerve-ending particles, Biochem. J., 97, 833 (1965).

9. R. J. A. Hosie, The localization of ATPase in morphologically characterized subcellular fractions of guinea-pig brain, Biochem. J., 96, 404 (1965).

10. R. W. Albers, R. Arnaiz, and E. DeRobertis, Na^+-K^+-ATPase and K^+-activated p-nitrophenylphosphatase: A comparison of their subcellular localizations in rat brain, Proc. Natl. Acad. Sci. U.S.A., 53, 557 (1965).

11. H. F. Bradford, E. K. Brownlow, and D. B. Gammack, The distribution of cation-stimulated ATPase in subcellular fractions from bovine cerebral cortex, J. Neurochem., 13, 1283 (1966).

12. A. A. Abdel-Latif, J. Brody, and H. Ramahi, Studies on Na^+-K^+-ATPase of the nerve endings and appearance of electrical activity in developing rat brain, J. Neurochem., 14, 1133 (1967).

13. J. Clausen and B. Formby, Comparative studies of K^+-p-nitrophenyl-phosphatase, K^+-acylphosphatase and Na^+-K^+-ATPase in synaptosomes of rat brain, Hoppe-Seyler's Z. Physiol. Chem., 349, 909 (1968).

14. A. A. Abdel-Latif, J. P. Smith, and N. Hedrick, Adenosinetriphosphatase and nucleotide metabolism in synaptosomes of rat brain, J. Neurochem., 17, 391 (1970).

15. T. Ohashi, S. Uchida, J. Nagai, and H. Hoshida, Studies on phosphate hydrolyzing activities in the synaptic membrane, J. Biochem., 67, 635 (1970).

16. T. D. White and P. Keen, Effects of inhibitors of Na^+-K^+-ATPase on the uptake of norepinephrine by synaptosomes, Mol. Pharmacol., 7, 40 (1971).

17. R. M. Marchbanks, The uptake of ^{14}C-choline into synaptosomes in vitro, Biochem. J., 110, 533 (1968).

18. L. T. Potter, "Uptake of Choline by Nerve Endings Isolated from the Rat Cerebral Cortex," in Interaction of Drugs and Subcellular Components in Animal Cells (P. N. Campbell, ed.), Little, Brown, Boston, 1968, p. 293.

19. I. Diamond and E. P. Kennedy, Carrier-mediated transport of choline into synaptic nerve endings, J. Biol. Chem., 244, 3258 (1969).

20. A. A. Abdel-Latif and J. P. Smith, Studies of choline transport and incorporation into lecithin of synaptosomes of delevoping rat brain, Brain Res., 31, 224 (1971).

21. C-M. Ling and A. A. Abdel-Latif, Studies on sodium transport in rat brain nerve-ending particles, J. Neurochem., 15, 721 (1968).

22. A. V. Escueta and S. H. Appel, Biochemical studies of synapses in vitro. II. Potassium transport, Biochemistry, 8, 725 (1969).

23. P. Keen and T. D. White, The permeability of pinched-off nerve endings to Na^+, K^+ and Cl^- and the effects of gramicidin, J. Neurochem., 18, 1097 (1971).

24. D. A. Grahame-Smith and A. G. Parfitt, Tryptophan transport across the synaptosomal membrane, J. Neurochem., 17, 1339 (1970).

25. R. W. Colburn, F. K. Goodwin, D. L. Murphy, W. E. Bunney, and J. M. Davis, Quantitative studies of norepinephrine uptake by synaptosomes, Biochem. Pharmacol., 17, 957 (1968).

26. T. D. White and P. Keen, The Role of internal and external Na^+ and K^+ on the uptake of [^{3}H] noradrenaline by synaptosomes prepared from rat brain, Biochim. Biophys. Acta, 196, 285 (1970).

27. A. S. Horn, J. T. Coyle, and S. H. Snyder, Catecholamine uptake by synaptosomes from rat brain, Mol. Pharmacol., 7, 66 (1971).

28. I. Diamond and A. L. Goldberg, Uptake and release of ^{45}Ca by brain microsomes, synaptosomes and synaptic vesicles, J. Neurochem., 18, 1419 (1971).

29. A. A. Abdel-Latif, A simple method for isolation of nerve-ending particles from rat brain, Biochim. Biophys. Acta, 121, 403 (1966).

30. E. D. Day, P. N. McMillan, D. D. Mickey, and S. H. Appel, Zonal centrifuge profiles of rat brain homogenates: Instability in sucrose, stability is iso-osmotic Ficoll-sucrose, Anal. Biochem., 39, 29 (1971).

31. M. K. Johnson and V. P. Whittaker, Lactate Dehydrogenase as a Cytoplasmic Marker in Brain, Biochem. J., 88, 404 (1963).

32. V. P. Whittaker, I. A. Michaelson, and R. J. A. Kirkland, The Separation of synaptic vesicles from nerve-ending particles ("Synaptosomes"), Biochem. J., 90, 293 (1964).

33. A. A. Abdel-Latif, J. P. Smith, and E. P. Ellington, Subcellular distribution of Na^+-K^+-ATPase, ACh, and AChase in developing rat brain, Brain Res., 18, 441 (1970).

34. E. DeRobertis, A. DeIraldi, G. Arnaiz, and L. Salganicoff, Cholinergic and non-cholinergic nerve endings in rat brain-I, J. Neurochem., 9, 23 (1962).

35. G. Dallner, P. Siekevitz, and A. E. Palade, Biogenesis of endoplasmic reticulum membranes. II. Synthesis of constitutive microsomal enzymes in developing rat hepatocyte, J. Cell Biol., 30, 97 (1966).

35a. A. A. Abdel-Latif and S. G. A. Alivisatos, Purification and properties of pyridine nucleosidase (Glycosidase) from bull semen, J. Biol. Chem., 237, 500 (1962).

35b. A. A. Abdel-Latif, M. Roberts, W. Karp, and J. P. Smith, Metabolism of phosphatidylcholine, phosphatidylinositol and palmitylcarnitine in synaptosomes from rat brain, J. Neurochem., 20, 189 (1973).

36. D. Nachmansohn and I. B. Wilson, "Choline Acetylase," in Methods in Enzymology, I (S. P. Colowick and N. O. Kaplan, (eds.), Academic, New York, 1955, p. 619.

37. P. D. Swanson, H. F. Bradford, and H. McIlwain, Stimulation and solubilization of Na^+-activated ATPase of cerebral microsomes by surface-active agents, especially poloxyethylene ethers: Actions of phospholipases and a neuraminidase, Biochem. J., 92, 235 (1964).

38. I. M. Glynn, C. W. Slayman, J. Eichberg, and R. M. C. Dawson, The ATPase system responsible for cation transport in electric organ, exclusion of phospholipids as intermediates, Biochem. J., 94, 692 (1965).

39. J. S. Charnock, A. S. Rosenthal, and R. L. Post, Studies of the mechanism of cation transport. II. A phosphorylated intermediate in the cation stimulated enzymic hydrolysis of ATPase, Australian J. Exptl. Biol. Med. Sci., 41, 675 (1963).

40. R. L. Post, A. K. Sen, and A. S. Rosenthal, A phosphorylated intermediate in ATP-dependent Na^+ and K^+ transport across kidney membranes, J. Biol. Chem., 240, 1437 (1965).

41. M. Pugita, T. Nakao, Y. Tashima, N. Mizuno, K. Nagano, and M. Nakao, K^+-Stimulated p-nitrophenylphosphatase activity occurring in a highly specific ATPase preparation from rabbit brain, Biochim. Biophys. Acta., 117, 42 (1966).

42. J. D. Judah, J. Ahmad, and A. E. M. McLean, Ion transport and phosphorproteins of human red cells, Biochim. Biophys. Acta., 65, 472 (1962).

43. Y. Israel and E. Titus, A comparison of microsomal (Na^+-K^+)-ATPase with K^+-acetylphosphatase, Biochim. Biophys. Acta., 139, 450 (1967).

44. R. Tanaka and T. Mitsumata, p-Nitrophenyl phosphatases of a membrane fraction from bovine cerebral cortex, J. Neurochem., 16, 1163 (1969).

45. B. Formby and J. Clausen, Phosphatase activity in particulate fractions of rat brain, Hoppe-Seyler's Z. Physiol. Chem., 349, 349 (1968).

46. L. E. Hokin, P. S. Sastry, P. R. Galsworthy, and A. Yoda, Evidence that a phosphorylated intermediate in a brain transport ATPase is an acyl phosphate, Proc. Natl. Acad. Sci. U.S.A., 54, 177 (1965).

47. A. L. Hodgkin and R. D. Keynes, Active transport of cations in giant axons from sepia and loligo, J. Physiol (London), 128, 28 (1955).

48. O. H. Lowry, N. J. Rosebrough, A. L. Farr, and R. J. Randall, Protein measurement with the folin phenol reagent, J. Biol. Chem., 193, 265 (1951).

49. A. Gomori, A modification of the colorimetric phosphorus determination for use with the photoelectric colorimeter, J. Lab. Clin. Med., 27, 955 (1942).

50. C. H. Fiske and Y. Subbarow, Colorimetric determination of phosphorus, J. Biol. Chem., 66, 375 (1925).

51. S. L. Bonting, "Na⁺-K⁺-activated ATPase and Cation Transport," in Membranes and Ion Transport (E. D. Bittar, ed.), Vol. 1, Wiley-Interscience, New York, 1970, p. 257.

52. A. A. Abdel-Latif, J. P. Smith, and C. A. Dasher, Rapid appearance of labeled lecithin and protein in isolated nerve endings from rat brain, Proc. Soc. Exptl. Biol. Med., 134, 850 (1970).

53. V. H. Ebel, J. R. Wolff, F. Dorn, and T. H. Günther, Wikung von Hormonen auf Elektrolytgehalt, ATPase and Endoplasmatisches Retikulum in Rattenhirn, Z. Klin. Chem. Klin. Biochem., 9, 249 (1971).

54. A. Atkinson, A. D. Gatenby, and A. G. Lowe, Subunit structure of the Na⁺, K⁺-dependent transport ATPase, Nature, 233, 145 (1971).

55. H. Matsui and A. Schwartz, Purification and properties of a highly active ouabain-sensitive Na⁺, K⁺-dependent ATPase from cardiac tissue, Biochim. Biophys. Acta, 128, 380 (1966).

56. A. Kahlenberg, N. C. Dulak, J. F. Dixon, P. R. Galsworthy, and L. E. Hokin, Studies on the characterization of the Na⁺-K⁺-ATPase V. Partial purification of the lubrol-solubilized beef brain enzyme, Arch. Biochem. Biophys., 131, 253 (1969).

57. R. F. Squires, On the interactions of Na^+, K^+, Mg^{++} and ATP with the Na^+-K^+-ATPase from rat brain, Biochem. Biophys. Res. Commun., 19, 27 (1965).

58. O. Jardetzky, Simple allosteric model for membrane pumps, Nature, 211, 969 (1966).

59. J. D. Robinson, Kinetic studies on a brain microsomal ATPase. evidence suggesting conformational changes, Biochemistry, 6, 3250 (1967).

60. R. Tanaka and L. G. Abood, Phospholipid requirement of Na^+, K^+-activated ATPase from rat brain, Arch. Biochem. Biophys., 108, 47 (1964).

61. R. Tanaka and K. P. Strickland, Role of phospholipid in the activation of Na^+-K^+-ATPase of beef brain, Arch. Biochem. Biophys., 111, 583 (1965).

62. M. Germain and P. Proulx, ATPase activity in synaptic vesicles of rat brain, Biochem. Pharmacol., 14, 1815 (1965).

63. L. G. Abood, E. Brunngraber, and M. Taylor, Glycolytic and oxidative phosphorylative studies with intact and disrupted rat brain mitochondria, J. Biol. Chem., 234, 1307 (1959).

64. V. P. Whittaker, The isolation and characterization of acetylcholine containing particles from brain, Biochem. J., 72, 694 (1959).

65. R. Tanaka and L. G. Abood, Isolation from rat brain of mitochondria devoid of glycolytic activity, J. Neurochem., 10, 571 (1963).

66. E. G. Brunngraber, J. Aguilar, and W. G. Occomy, The intracellular distribution of glycolytic and tricarboxylic acid cycle enzymes in rat brain mitochondrial preparations, J. Neurochem., 10, 433 (1963).

67. A. A. Abdel-Latif and L. G. Abood, Biochemical studies on mitochondria and other cytoplasmic fractions of developing rat brain, J. Neurochem., 11, 9 (1964).

68. S. H. Appel and B. L. Parrot, Hexose monophosphate pathway in synapses, J. Neurochem., 17, 1619 (1970).

69. D. J. K. Balfour and J. C. Gilbert, Studies of the respiration of synaptosomes, Biochem. Pharmacol., 20, 1151 (1971).

70. H. F. Bradford and A. J. Thomas, Metabolism of glucose and glutamate by synaptosomes from mammalian cerebral cortex, J. Neurochem., 16, 1495 (1969).

71. A. A. Abdel-Latif and L. G. Abood, Incorporation of ^{32}P into the subcellular fractions of the developing rat brain, J. Neurochem., 12, 157 (1965).

72. A. A. Abdel-Latif, Acetylcholine and the incorporation of ^{32}P into phospholipids and phosphoproteins of nerve endings of developing rat brain, Nature (London), 211, 530 (1966).

73. J. Durell and M. A. Sodd, Studies on the acetylcholine-stimulated incorporation of ^{32}P into the phospholipid of brain particulate preparations. II, J. Neurochem., 13, 487 (1966).

74. A. A. Abdel-Latif and J. P. Smith, In vitro incorporation of ^{14}C-choline into phosphatidylcholine of rat brain synaptosomes and the effect of calcium ions, Biochem. Pharmacol., 21, 436 (1972).

75. I. G. Morgan and L. Austin, Synaptosomal protein synthesis in a cell-free system, J. Neurochem., 15, 41 (1968).

76. L. A. Autilio, S. H. Appel, P. Pettis, and P. Gambetti, Biochemical studies of synapses in vitro, Biochemistry, 7, 2615 (1968).

77. H. F. Bradford, Metabolic response of synaptosomes to electrical stimulation: Release of amino acids, Brain Res., 19, 239 (1970).

78. E. G. Gray and V. P. Whittaker, The isolation of nerve endings from brain: An electron-microscopic study of cell fragments derived by homogenization and centrifugation, J. Anat. (London), 96, 79 (1962).

79. R. M. Marchbanks, The osmotically sensitive K^+ and Na^+ compartments of synaptosomes, Biochem. J., 104, 148 (1967).

80. A. A. Abdel-Latif, M. Yamaguchi, J. Smith, and T. Yamaguchi, Studies on the effect of ouabain on sodium and phosphate uptake into nerve endings of developing rat brain, Life Sci., 7, 1325 (1968).

81. A. A. Abdel-Latif and J. P. Smith, Studies on choline transport and metabolism in rat brain synaptosomes, Federation Proc., 31 (1972).

82. L. T. Potter, Uptake of propranolol by isolated guinea-pig atria, J. Pharmacol. Exptl. Therap., 155, 91 (1967).

83. P. Keen and T. D. White, A light-scattering technique for the study of permeability of rat brain synaptosomes in vitro, J. Neurochem., 17, 565 (1970).

84. A. L. Koch, Some calculations on the turbidity of mitochondria and bacteria, Biochim. Biophys. Acta, 51, 429 (1961).

85. P. G. LeFevre and M. E. LeFerve, The mechanism of glucose transfer into and out of the human red cell, J. Gen. Physiol., 35, 891 (1952).

86. H. Tedeschi and D. L. Harris, The osmotic behavior and permeability to non-electrolytes of mitochondria, Arch. Biochem. Biophys., 58, 52 (1955).

87. P. Keen, personal communication.

88. R. M. Marchbanks, Compartmentation of acetylcholine in synaptosomes, Biochem. Pharmacol., 16, 921 (1967).

89. J. R. Cooper, F. E. Bloom, and R. H. Roth, The Biochemical Basis of Neuropharmacology, Oxford University, New York, 1970.

90. R. W. Ryall, The subcellular distribution of acetylcholine, substance P, 5-hydroxytryptamine, γ-aminobutyric acid and glutamic acid in brain homogenates, J. Neurochem., 11, 131 (1964).

91. H. F. Bradford, Respiration in vitro of synaptosomes from mammalian cerebral cortex, J. Neurochem., 16, 675 (1969).

92. C-P. Sung and R. M. Johnstone, Evidence for active transport of choline in rat kidney cortex slices, Can. J. Biochem., 43, 1111 (1965).

93. A. L. Hodgkin and K. Martin, Choline uptake by giant axons of loligo, J. Physiol. (London), 179, 26 P (1967).

94. J. Schuberth, A. Sundwall, B. Sorbo, and J-O Lindell, Uptake of choline by mouse brain slices, J. Neurochem., 13, 347 (1966).

95. J. Schuberth, A. Sundwall, and B. Sorbo, Relation between Na^+-K^+ transport and the uptake of choline by brain slices, Life Sci., 6, 293 (1967).

96. W. J. Cooke and J. D. Robinson, Factors influencing choline movements in rat brain slices, Biochem. Pharmacol., 20, 2355 (1971).

97. A. Askari, Uptake of some quaternary ammonium ions by human erythrocytes, J. Gen. Physiol., 49, 1147 (1966).

98. J. Martin, Concentrative accumulation of choline by human erythrocytes, J. Gen. Physiol., 51, 497 (1968).

98a. A. A. Abdel-Latif and J. P. Smith, Studies on choline transport and metabolism in rat brain synaptosomes, Biochem. Pharmacol., 21, 3005 (1972).

98b. B. C. Pressman, E. J. Harris, W. S. Jagger, and J. H. Johnson, Antibiotic-mediated transport of alkali ions across lipids barriers, Proc. Nat. Acad. Sci. U.S.A., 58, 1949 (1967).

99. Y. Itokawa and J. R. Cooper, Ion movements and thiamine, II. The release of the vitamin from membrane fragments, Biochim. Biophys. Acta, 196, 274 (1970).

100. A. Lehninger, Biochemistry, Worth, New York, 1970, p. 605.

101. R. Whittam and K. P. Wheeler, Transport across cell membranes, Ann. Rev. Physiol., 32, 21 (1970).

102. D. F. Bogdanski, A. Tissari, and B. B. Brodie, Role of Na$^+$, K$^+$, ouabain and reserpine in uptake, storage and metabolism of biogenic amines in synaptosomes, Life Sci., 7, 419 (1968).

Chapter 5

CRITICAL EVALUATION AND IMPLICATIONS OF DENERVATION AND
REINNERVATION STUDIES OF CROSS-STRIATED MUSCLE

E. Gutmann
Institute of Physiology
Czechoslovak Academy of Sciences
Prague, Czechoslovakia

I. INTRODUCTION

A. Neuron-Muscle Relationships

Complex relationships exist between nerve and muscle cells, which
have been extensively studied in growth and development (1, 2), establish-
ment of connections in tissue cultures (3-5), regeneration of limbs in lower
vertebrates (6), and in denervation and reinnervation studies, especially
in mammals (7-9). The latter studies concerned the changes following interruption and reestablishment of the communications between nerve and

189

muscle cells and have conclusively shown a high metabolic interdependence between muscle and nerve cells. However, they have also demonstrated that transmission of excitation across the neuromuscular synapse and the related liberation of the neurotransmitter reflect only one aspect of this complex relationship (10-13) and that other factors are also involved.

Studies of electrical recordings from endplates and neuronal synapses have shown that the neurotransmitter is secreted spontaneously in minimal "quantal" packets, discharged at a much higher frequency during functional activity, with ensuring alterations of the postsynatpic membrane (14). The complementary biochemical and ultrastructural evidence suggested that the transmitter is parcelled up in synaptic vesicles and that consequently the synthesis and release of the transmitter has often been thought sufficient to explain the whole mechanism of nerve-muscle cell communication and related metabolic reactions (15). Release of the neurotransmitter connected with (a) nerve impulse activity, (b) spontaneous "secretory" (i.e., impulse-independent) effect, or (c) release of neurotrophic agents, so far chemically undefined, might be responsible for the trophic effects of neuronal activity.

The concept that the dependence of muscle on nerve cannot be solely explained by neuronal activity related to transmitter release has been strengthened by observations dealing with

(a) Synthesis, transport, and release of both proteins and neurotransmitters (16-18),

(b) General mechanisms of neurosecretion, i.e., secretion of diverse active proteins in the neuron (19),

(c) The phenomenon of axoplasmic transport was first studied in regenerating nerves (20, 21) after nerve constriction (22); it led to a concept of transport of proteins moving with widely varyings rates (23, 24).

(d) The complex nature of the synaptic apparatus. The latter involves macromolecular systems in the synthesis of vesicles and surface recognition (16) and suggested mechanisms of information transfer between cells beyond transmitter-produced alterations in excitatory states.

(e) Recognition of a metabolic effect of the distal nerve stump after nerve section on muscle independent of nerve impulse transmission. After nerve section, onset of the rate of break-down of endplates and stability of the muscle membrane (25, 26), development of hypersensitivity to ACh, failure of neuromuscular transmission (25, 27) as well as abnormal metabolism (28) all depend on the length of axon. These observations supplied strong evidence for the significance of axoplasmic transport and effects on the postsynaptic cell, not connected with nerve impulse activity and neuromuscular transmission.

(f) Apparently nonspecific growth-promoting influence of nerves in amphibian limb regeneration, independent of impulse conduction (29, 30).

(g) Some specific, morphological, physiological, and biochemical changes in muscle after nerve section, known as denervation atrophy, which affect the stability of membrane and intracellular constituents (8).

(h) Nonspecific "trophic" effects of neurons on muscle in tissue cultures without synaptic transmission (31) and even without contact between nerve and muscle cell (32), found in neuroblastoma (31) or sensory ganglia (32) cells.

These and other experiments suggest, that neither impulse-directed nor spontaneous transmitter (acetylcholine) release can be the only or decisive mediator of neurotrophic effects.

B. Functional-Metabolic Correlations

Nerve section evidently leads to a progressive loss of maintenance of both the neuron and innervated tissue, as demonstrated by morphological, functional, and biochemical changes in the nerve cell and degeneration of the peripheral cell process, accompanied by atrophy and degeneration of muscle fibers. Profound changes in the regulation of intercellular relationships are apparently initiated by nerve section and are reversed by the regeneration process (8). Nerve section thus shows the mechanism by which neuron and muscle cells ("center" and "periphery") interact. "Periphery" can be defined as the "set of postsynaptic cells for the system in question" (33). We are thus confronted with analogous problems and questions concerning both central and peripheral nervous system. Section of central or peripheral axons leads to changes of both post- and presynaptic cells, apparently by a mechanism of "double dependence" (34).

Three basic neurohistological observations after nerve section gave the first clues for the rapid protein synthesis in the perikaryon and the regulation of intercellular relations disturbed after nerve section (Fig. 1), i.e., (a) Nissl's (35) "primary cell irritation" or chromatolysis in the perikaryon of the neuron, (b) Cajal's (20) observations on growth cones, and (c) the "attempt at abortive regeneration" at the ends of the peripheral stump and apparently also at central axon terminals (36) preceding final degeneration. The first observation indicated an increase of protein synthesis in the nerve cell after nerve section, and dispersion of "Nissl substance" (37), which corresponds to RNA attached to the endoplasmic reticulum (38). The other findings show the importance of convection of proteins [axoplasmic flow (22)] in regeneration experiments. These observations are in line with the concept of the neuron as a general secretory cell, not only restricted to neurohormonal messengers (19).

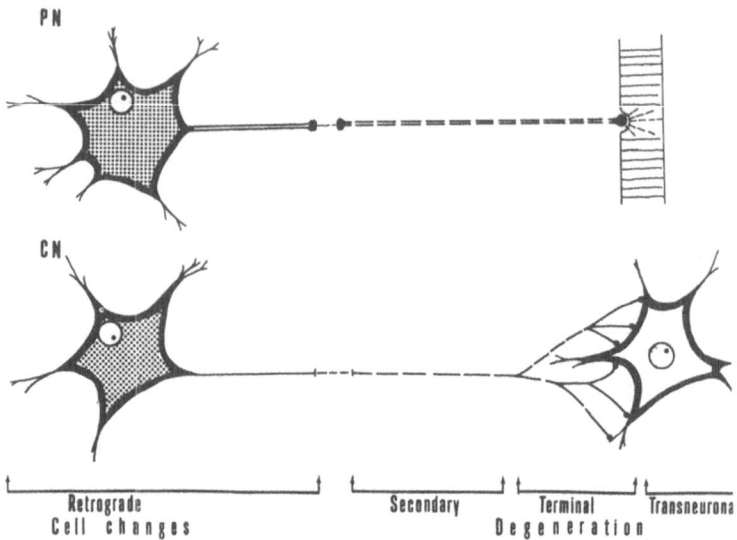

FIG. 1. Scheme of the changes taking place in neurons after reaction of peripheral nerves (PN) and of central pathways (CN) indicating "Chromatolysis" (35), in cell body and "attempts at abortive regeneration" (20).

The concept of the secretory function of the neuron facilitated the study of the cellular dynamics of the neuron, not restricted on to transmission of the nerve impulse and synthesis, transport and release of neurotransmitters, i.e., ACh, in nerve-muscle systems. Thus, increase of protein synthesis will occur after section of ventral roots in motor neurons, or in dorsal ganglion cells after section of the peripheral cell processes, apparently independent of the direction of nerve impulses (39). Denervation of the receptor organs is often followed by degeneration of the sensory cell (1); sensory neurons transport newly synthesized proteins at a rapid rate away from the cell body, i.e., also in the direction opposite to conduction of the nerve impulse (40).

Neuronal secretion of proteins is apparently not necessarily related to neurotransmitter release. Secretion of "trophic" substances, possibly proteins, seems to be involved in nerve-dependent regeneration of the newt limb (41). Blocking of neuromuscular transmission does not reproduce denervation changes (8). During reinnervation of muscle, there is a delay between recovery of neuromuscular contact and neuromuscular transmission (42). Electrical membrane properties (43), sensitivity to ACh (10) (which is normally restricted to the synaptic region but affects the whole muscle membrane after denervation), and also the contracture response of muscle after cross-unions of nerve (44) change before recovery of neuromuscular transmission. In tissue culture, early formation of synapses with high sensitivity to ACh only at the contact region appears to be also established before functional synaptic neuromuscular transmission (31).

The alterations in membrane stability (e.g., fibrillation) (45), in sensitivity to ACh (46), and in intracellular metabolism show some specific features which cannot be explained by loss of impulse-directed transmitter release only. Activities of the neuron not related to the nerve impulse have been defined as trophic (8). However, spontaneous transmitter release alone also cannot explain all neurotrophic relationships (9, 39). The neuron apparently does not only secrete a transmitter substance but also proteins whose biological activities may be no less specific than those of the transmitter (17). It must, however, be admitted that these proteins have not yet been chemically defined and that the mechanisms of transfer of such proteins from neuron to muscle (32, 47, 48) remain unclear.

C. Neurochemical Implications and Significance

Studies on communications between nerve and muscle cell, especially denervation studies, should have considerable neurochemical interest as they provoke questions and supply information on (a) the nature of intracellular relations between neuron and innervated tissues and between neurons themselves, (b) functional-metabolic correlations in neuronal activities affecting innervated tissues, and (c) specifity of neuronal function and metabolism.

Denervation, which studies changes after nerve section, of course, is limited because it is essentially an "elimination" experiment (13), but it does supply potent evidence for evaluation of the effect of disuse on neuronal function and indirectly on mechanisms operating in use and disuse. Cross-union experiments of nerves, which study the results of change of nerve supply, i.e., innervation of fast and slow muscles by "foreign" nerves, have shown a change of speed of muscular contraction (49), enzymic properties of myosin (50), of metabolic (51) and enzymic (52) properties of muscle (52), and of histochemistry of muscle fibers (53, 54) and have brought new emphasis to the issue of specifity of neuronal metabolism. These studies have neurogenetic implications and suggest neural regulation of gene expression in muscle (55).

Nerve section is a simple procedure and reinnervation and cross-innervation experiments are also relatively easy to perform. This simplicity has often led to underestimation of experimental artifacts which might be misleading, especially when such experiments lead to neurogenetic generalizations. A section on methods of denervation and reinnervation and on the limitations and contributions of such studies appears therefore to be justified before a detailed discussion of problems of differentiation of muscles, their specifity, and the effects produced by changing their innervation.

II. METHODS

A. Surgical Methods

The changes in neuron and end organ following nerve section are bio-
logically closely related to the regeneration process (8). This is indicated
by the initial increase of protein synthesis in the perikaryon of the nerve
cell, the "primäre Reizung" of Nissl (35), and the peripheral nerve stump
due to increased mitotic activity in the Schwann cells (56, 57) and a consequent decrease of these activities if reinnervation of the peripheral nerve
stump does not take place. Recovery of normal intercellular relationships
follows the regeneration of the axons and the reinnervation of the end organ,
whereas lack of regeneration leads to progressive and marked atrophy of
nerve cells and peripheral nerve stump, resulting in fibrotic "Büngner
bands."

A denervation experiment aims at complete interruption of the nerve
and should avoid reinnervation and possible side effects, whereas reinnervation experiments have to take into account these intercellular relationships, i.e., the changes in neurons and Schwann cells which depend on the
time of reinnervation after nerve section (7). Surgical methods of nerve
section and repair, initiating reeinnervation of muscle, are now considered.

1. Nerve Section

Clinical experiences have shown that the nature, site, and severity
of nerve injury introduce complicating factors and have led fundamentally
to the distinction of three types of nerve injury (58), two of them easily
reproducible in experimental work.

(a) Temporary, i.e., reversible interruption of conduction without
loss of axonal continuity between neuron and end organ, i.e., nerve block
produced for instance by infusion of 2% novocaine in the nerve (59) or sili-
cone polymer-lidocaine implants around the nerve which eliminates rever-
sibly nerve impulses up to 14 days (60). Short nerve blocks of poor reproducibility can also be evoked by compression (61) or ischaemia (62).
Block of neuromuscular transmission can be produced by curare; however,
respiratory problems (especially in rats) and pharmacological side effects
limit use of this approach.

(b) Complete nerve section affecting interruption of axons, endo-
perineurium, and connective tissue wall.

(c) Nerve crush leads to degeneration of the axons below the lesion
with preservation of the general structure of the endoneurial sheaths.
This kind of lesion allows fast reinnervation and provides optimal

experimental conditions for regeneration experiments if the crush lesion is performed at a single point.

In denervation experiments, we are primarily concerned with nerve section that ensures optimal reproducibility of results. Pharmacological denervation, e.g., by intramuscular application of botulin toxin (15, 63) or mercaine, is difficult to reproduce. Mercaine (64) is applied subcutaneously and therefore affects only the superficial part of the muscle. Interpretations of the results are therefore difficult unless single muscle fibers are studied and denervation tested by intracellular registration. Moreover, denervation changes may be due to direct myotoxic effects.

Nerve section is a quick procedure and can best be done under ether anesthesia, unless short-term changes in enzymatic pattern of muscle are studied. In prolonged, time-consuming experiments, barbiturates, applied intravenously or intramuscularly, are to be preferred. Nembutal (0.6 ml/kg body weight) is recommended. No aseptical procedures are necessary for rats and mice, but they are necessary for experiments on rabbits, cats, and dogs. Denervation experiments in rats do not need assistance and may be finished in 5 min. Experiments on rabbits or cats need one assistant and are more time consuming (15-20 min). Shaving the skin, incision of skin and fascia and retraction of overlying muscles is followed by freeing the nerve trunk from the surrounding fascia. The nerve is lifted with a thin glass hook at one point only and cut at one point perpendicular to the nerve stump with fine sharp scissors. Compression and traction especially of the central nerve stump is to be avoided. Since tensile strength and elastic properties of peripheral nerves depend on the perineurium, stretching of the nerve may damage neuronal function. Experiments on nerve roots which lack tough protective epineurinal and perineurial sheaths require greater care. A precisely defined site of lesion is a prerequisite for comparability and reproducibility of denervation experiments. In short-term experiments the distance from lesion to muscle should be stated.

Spinal nerves from several segments of the spinal cord participate in plexus formations in which motor and sensory fibers from different segments of the cord and sympathetic fibers are intermingled and rearranged. Thus, peripheral nerves contain motor and sensory fibers from a number of cord segments. The localization of the corresponding neurons is necessary, especially when the effect of changes in muscle on nerve cells are studied. The peripheral nerves, most used in denervation experiments, i.e., peroneal (N. peroneus) and tibial (N. tibialis) nerve, have their spinal neurons in the rat in L_4-L_5 (65) and in L_6-S_1 in the cat (66). Moreover, branching of fibers in the nerve trunk must be taken into consideration. For instance, individual nerve fibers branch while coursing in human

cutaneous nerve trunks (58). It should be remembered, that peripheral
nerves are composed of several funiculi which engage in plexus formations
along the full length of the nerve (58). Regrouping of fibers take place,
fibers come to occupy different funiculi with nerve fibers collected for
specific cutaneous and muscle branches. The resulting change in cross-
sectional area of nerves (Fig. 2) has to be kept in mind in both denervation
and reinnervation studies. Variations in intraneural topography are in-
evitable, but best reproducibility is achieved with nerve section before
entry of the nerve into muscle; such a section achieves relatively uniform
cross-sectional area of nerve and avoids trophic cutaneous disturbances
resulting from section of skin nerves. Trophic ulcers due to denervation
of skin area can be avoided, at least to a great extent, even after section-
ing the whole sciatic nerve, if the sural nerve supplying a large cutaneous
area is spared (67). The sural nerve is dissected free from the tibial nerve
nerve in the middle of thigh and the sciatic nerve is then cut just below the
highest point of dissection. Incidence and severity of trophic skin disturbances (ranging from dryness and epidermal atrophy to edema and ulcers)
is usually more marked after high lesions. This is due to more extensive
vasomotor paralysis, greater area of sensory denervation, and more
severe retrograde neuronal changes. Trauma and pressure to the in-
sensitive skin are the most important factors producing ulceration. Such
severe trophic disturbances are absent after section of the nerve before
entry into the muscle. Section of the peroneal nerve, due to a relatively
small contribution of sensory nerve fibers and absence of pressure to the
denervated skin, leads to small trophic disturbances only (58).

It should be noted that in all experiments the time of denervation should
be exactly stated, as the time gradient of metabolic and functional changes
may differ in different muscles at different periods of denervation (68).

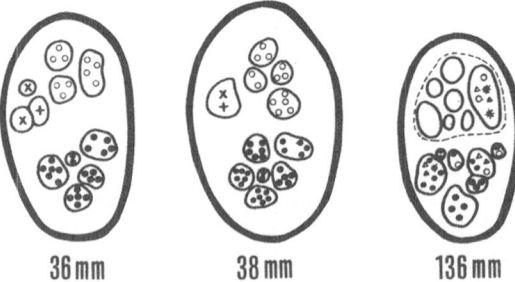

FIG. 2. Transverse sections of the human radial nerve 36, 38, and
136 mm above the lateral humeral epicondyle, illustrating the change in
funicular pattern in a mixed nerve trunk; ● posterior interossens, ▲ brach-
ialis, o superficial radial, x extensor carpi radialis longus, + extensor
carpi radialis crevis, * combined radial extensors of the wrist (reprinted
from Ref. 58, p. 205).

2. Nerve Repair

The aim of nerve repair, from a neurosurgical point of view, is to bring about reinnervation of muscle and full recovery of function. From a neurochemical point of view it affords an adequate model for studying the recovery of metabolism of the muscle cell due to neuronal activity and the restoration of metabolic nerve-muscle cell relations. Recovery will depend on the degree of success of different components involved in nerve regeneration: (a) the recovery of the neuron from retrograde effects, (b) growth of axon tips up to and across the site of lesion and along the endoneurial tubes, (c) maturation of the axons, (d) formation of connections with the muscle cell, and (e) the recovery of normal metabolism and function of muscle (2, 7, 21, 69).

(a) Nerve Crush. Full recovery can be expected only after lesions by nerve block and nerve crush. Crushing the nerve with a fine watchmaker's forceps at one point only leaves the continuity of the endoneurial tubes intact and results in full reconstitution of the nerve in respect to number and diameter of nerve fibers. The sciatic nerve has been used in most of the experimental work. Crushing a single nerve is, however, preferable. Distance of nerve crush from muscle or nerve cell respectively should be noted, especially in short-term experiments. The peroneal nerve, which innervates fast muscles, or the soleus nerve, which innervates the slow soleus muscle, can be crushed just before their entry into the muscles. If the peroneal nerve is crushed higher up, it has to be dissected free from the tibial nerve using a sharp scalpel. The nerve is then lifted carefully with a thin glass hook and crushed with a watchmaker's forceps at one point (Fig. 3).

Maturation of nerve fibers is a prolonged process; the number, size, and size pattern of nerve fibers in the rabbit is fully reconstituted only after 250-300 days (70) and conduction velocity after 456 days (71) below a crush lesion in the rabbit. Essentially, the degree of myelinization is determined by the nature of the parent cell (72) but will also depend on whether and when connection with the end organ is accomplished. Delay of nerve regeneration — and consequently of reinnervation of the muscle — will decrease the degree of reversibility of denervation changes in muscle (73).

A delay or reinnervation after a nerve crush, the method of choice when full muscle recovery is required, can be achieved by repeated crushing of the nerve (73, 74) and reveals an almost unlimited capacity of regeneration of axons in the adult. Any interruption of the continuity of endoneurial tubes (i.e., after nerve section) will reduce the degree of successful regeneration due to a multitude of factors, such as increase of latency periods (20), rate of regeneration in the peripheral stump, and greater

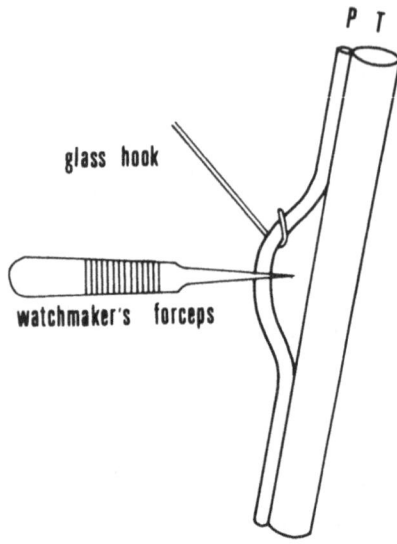

FIG. 3. Scheme of procedure in nerve crush. The peroneal nerve is dissected free from the tibial nerve, lifted with a thin glass hook, and crushed at one point with a watchmaker's forceps. Stretching or rotation of nerve is avoided.

chance of formation of inadequate connections (58). Axons have the capacity to bridge long gaps in injured nerves, but the capacity of successful regeneration is progressively reduced especially by length of the gap and delay of nerve repair. Thus, the method of localized nerve crush is optimal for study of recovery of metabolic relationships between nerve and muscle cell.

(b) Nerve Suture. Nerve sutures require mobilization of the nerves to permit transposition of the nerves to a new bed and joining separated ends without tension. Tension at the suture line and unnecessary trimming of nerve ends is to be avoided as this makes proper apposition of the nerve ends difficult and increases the change in cross-sectional area. No ideal similarity in cross-sectional area of central and peripheral nerve stumps can be expected because of the intraneuronal shifts of funiculi (58). In end-to-end suture of the same nerves, oblique section, unnecessary tension, rotation of the nerve stumps, and the use of large nerve trunks containing very different nerve fibers should be avoided because they result in an increase of connections of functionally unrelated funicular groups and diminish success of regeneration.

Primary (i.e., immediate) is preferable to a secondary (i.e., delayed) nerve suture due to increasing funicular atrophy following disintegration and removal of degenerated axons and myelin. This atrophy increases with the duration of denervation, especially up to three months after nerve section; the resulting skrinkage of the cross sectional area of the distal nerve stump will make adequate apposition of the stumps impossible. The suture material should set up a minimal fibroblastic reaction. Fine silk or human hair is the most satisfactory suture material and the use of so-called "atraumatic" needles (Ethicon Silk 7-0) is advisable. The stumps are adjusted with a fine watchmaker's forceps and sutured epineurally. Suturing thicker nerves in larger animals may need two lateral sutures (Fig. 4). In the rat only one suture with the thin silk suture material is applied. Use of plain catgut or catgut and tantalum is not advisable in experimental work. Sutureless nerve union (including plasma clot, fibrin adhesive, the micropore adhesive type, and the arterial sleeve method) is advisable especially when thin nerves are used.

The plasma clot method (75) uses cockerel blood plasma fortified before application with fibrinogen and embryo tissue extract which then coagulates at the suture line. This method gives very good comparable results, as evidenced by the linear rate of regeneration of axons (76). This method is, of course, not suitable when tension exists at the suture line but should be preferable to conventional epineurial thread suture (77). The fibrin adhesive method (78) uses a smooth film of fibrin, prepared from fractionated human blood impregnated with thrombin and coated with fibrinogen on one side, which will be applied to the nerves. The junctional region may also be enclosed in a tubular film of fibrin. A high strength of

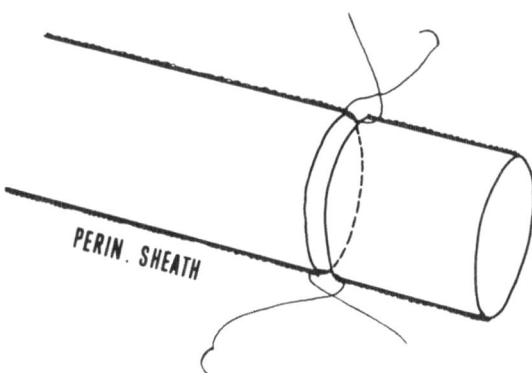

FIG. 4. Scheme of nerve suture. The perineural sheath is lifted with the tip of a fine watchmaker's forceps and two thin silk sutures are applied through the perineural sheath joining proximal and distal stump of the nerve at two lateral points.

union has been claimed. No advantages are introduced by the micropore adhesive tube method (79) or the tubular splicing method with arterial sleeves (80).

Adequate apposition and relative equality of size of nerve ends are very important for proper interpretations and the successful regeneration, especially in nerve cross-union experiments, where a priori a different funicular pattern must be expected. Union of the large tibial with the small peroneal nerve will of course produce different changes for reinnervation (Fig. 5). However, not only qualitative factors affecting the number of axons admitted to the peripheral nerve trunk and end organ are operative. Growth of the nerve fibers, which for a long time was thought to be a random process, nonselective, and subject only to mechanical guidance ["contact guidance" (81)], appears to be affected by selective forces originally termed "neurotropic" (20). The factors related to a selective preferential growth during development and regeneration (82) are still poorly understood (2); however, evidence for their importance is growing (83, 84) and these mechanisms must be taken into account especially in choice of adequate cross-innervation experiments.

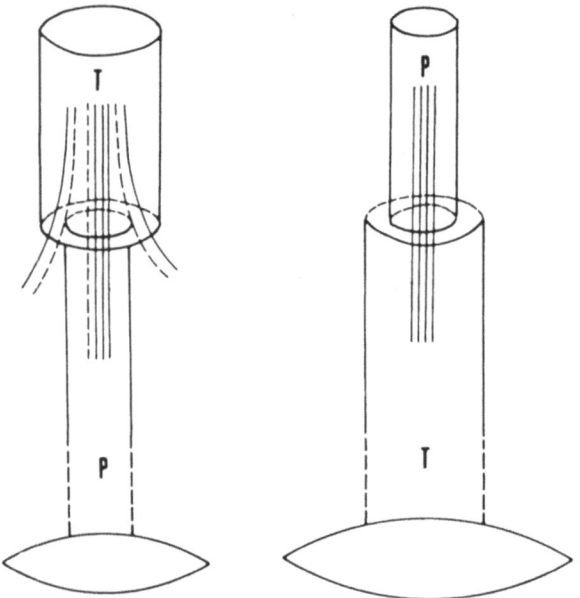

FIG. 5. Scheme of experimental conditions in cross-union experiments of tibial into peroneal (left) and peroneal into tibial (right) nerve indicating the greater chance of loss of slow type fibers (interrupted lines) into the scar in tibial-peroneal nerve cross-unions.

(c) Heteroinnervation. "Hetero-innervation," i.e., innervation of muscle by "foreign" nerve supply, is essentially an experimental test of specifity of the neuron and plasticity of muscle induced by the neuron. Muscles and muscle fibers differ in functional, biochemical, and morphological properties and achieve and maintain this differentiation, essentially of "fast" and "slow" muscle fibers during development (85-87), the neuron being of basic importance in these processes (88, 89). Thus nerve cells and the muscle fibers they innervate are matched by analogous properties. This is due to the fact that muscle fibers which make up a given skeletal muscle are organized into functional entities, the motor units, i.e., the population of muscle fibers innervated by a single motoneuron (90). Muscles contain two to three types of muscle fibers, contracting with fast, intermediate, and slow velocity. They are also distinguished histochemically according to ATPase activity (91); type II fibers have a high ATPase activity, type I fibers a low activity or an inverse relationship between oxidative and glycolytic enzymes (92). The classification of A, B, and C fibers concerns succinic dehydrogenase activity (SDH). A fibers have low activity while C fibers have high activity (93). More refined differentiation of muscle fibers has been undertaken on an ultrastructural basis (94, 95) and further ultrastructural and microelectrophysiological studies will, in the future, most likely suggest more detailed classifications. The heterogeneity of muscle is, however, basically the result of the intermingling of two to three different distinct motor units (96-98) in which nerve and muscle cells are "typed" in relation to physiological, structural, and metabolic characteristics. Speed of contraction and the closely related ATPase activity is the most consistent basis for differentiation; accordingly, the differentiation of "fast" and "slow" muscles (i.e., according to prevalence of fast or slow motor units) and muscle fibers is preferable. "Fast" and "slow" motoneurons differ, for example, in size, excitability, and other parameters (99, 100) and innervate fast and slow muscles. These terms are used here for the sake of basic importance and simplicity.

"Heteroinnervation" experiments should therefore reply to the questions whether or to what extent neural and/or myogenic influences determine the properties of different, i.e., fast and slow muscle fibers.

"Heteroinnervation," i.e., reinnervation of a muscle by a "foreign" nerve supply, can be achieved by (a) nerve implantation and (b) cross-union of nerves. Nerve implantation can be performed only into a denervated muscle, since a normally innervated muscle in adult animals rejects an additional nerve supply (101). However, implantation into an innervated muscle, i.e., an additional nerve supply is possible by the use of preimplantation (102) or regeneration (103), making use of a tempory denervation period. In ordinary implantation experiments, the original ("old") nerve is sectioned and ligated and the "new" or "foreign" nerve is cut distally before entry into the "old" muscle, to afford a sufficiently long nerve stump and is joined to the muscle with a thin silk suture (Fig. 6). In some cases, a ligated sensory branch can be used for guidance and fixation of the motor nerve (103). Use is made also of fibrin clots to fasten the nerve into a small incision of the fascia on the muscle. In adult animals implantation (heteroinnervation) with formation of new endplates is possible only after surgical (104) or pharmacological (105) denervation, unless local muscle fiber injury has occurred (106). The implanation method has the disadvantage that only a small, mostly superficial part of the muscle can be heteroinnervated. The method is therefore more suitable for single muscle fiber studies than for studies of the whole muscle.

Cross-unions enable innervation of the whole muscle by a foreign neuron. The general considerations of nerve sutures apply also to cross-unions; however, the nerve sutures require considerable mobilization of the nerves to avoid tension at the suture lines and to allow free transposition of the nerves to a new bed. The proximal ("old") nerve stump (e.g., of the N soleus) must be ligated to make "self-innervation" impossible. The new "foreign" (e.g., the N. peroneus) nerve is cut as distally as possible and sutured to the old peripheral end of the "old" nerve (Fig. 7). Thus the main technical tasks in cross-union experiments are the avoidance of self-reinnervation and the availability of long stretches of the new central and old peripheral stumps of the nerves to be "cross-united."

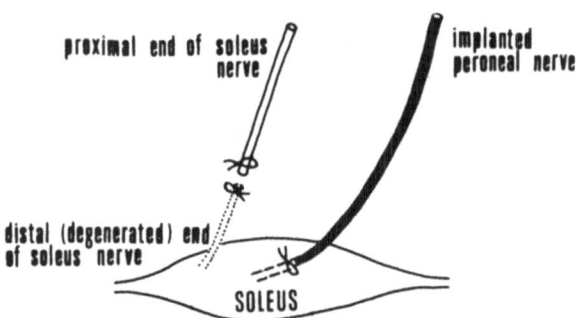

FIG. 6. Heteroinnervation of the soleus muscle by the peroneal nerve (nerve-implantation). The nerve to the soleus muscle is cut and a ligature is applied to its proximal and distal end. The peroneal nerve is cut near its entry into the muscles, dissected free, mobilized, and attached to the soleus muscle. An incisure is made through the fascia of the soleus muscle, and the peroneal nerve is fixed under the fascia. Fixation of the peroneal nerve to the soleus muscle is ensured by one-two ligatures through the fascia.

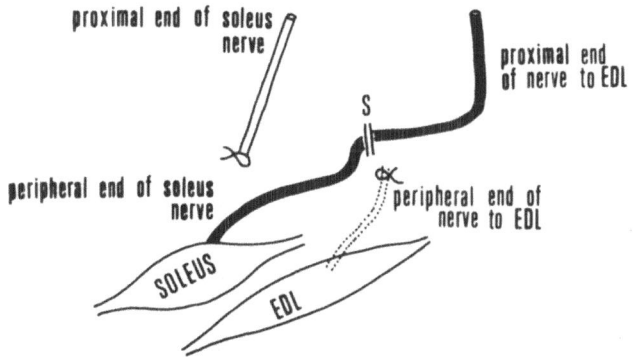

FIG. 7. Scheme of heteroinnervation of the soleus muscle by cross-union of nerve to extensor digitorum muscle (EDL) and nerve to soleus muscle. The soleus nerve is cut as distally from the muscle as possible, and its proximal end is ligated. The nerve to the EDL or the peroneal nerve is cut near to entry into the muscle and a ligature is applied to its peripheral end. Both proximal end of nerve to EDL (or of the peroneal nerve, which supplies fast muscles only) and the peripheral end of the soleus nerve are freed from the fascia, mobilized, and after transposition into a new bed, sutured with a thin suture material or joined by a plasma clot.

B. Choice of Animal

Choice of animal is dictated by those aspects on which regeneration studies are centered, but certain general considerations have to be made.

In studies on the effect of nerve section on neuronal function and metabolism data on localization of various motor cell groups innervating specific muscles should be consulted, such as those of Romanes (107) for the cat and rabbit. In the frog, the large motor neurons are not grouped into distinct masses of cells as in mammals (2). Knowledge of the spinal segments supplying the nerves in question is important. In the rat, the experimental animal most frequently used in denervation studies, the ischiadicus nerve is supplied from L_4-S_1, and the related position of the vertebrate can easily be located (65). Age is an important factor since nerve section at early stages of development leads to irreversible damage of a large number of motor neurons. The chromatolytic reaction in nerve cells used to map the motor cell groups in the spinal cord is easily produced only in very young mammals (107).

In denervation and reinnervation studies of muscles, basic data allowing functional correlations should be known. Data on the composition of peripheral nerves (for the cat see 108), on rate of regeneration (23, 76), and rate of muscle atrophy (109, 110) in different animals should be consulted. Data on loss of neuromuscular transmission and of spontaneous transmitter release are important. These differ in different animals. For instance, loss of neuromuscular transmission after nerve section occurs after 10-20 days in cold-blooded vertebrates and after 32 hr in the rabbit. It is very helpful if these data are tabulated (23) but the exact level of the lesion, age, and method of testing are not always known, and these factors make comparisons difficult.

For the study of neuromuscular relationships isolated nerve-muscle cell preparations (organ cultures) in vitro should be ideal. Such approaches were used, for instance, in the snail (47) and frog (111). Nerve-muscle preparations of the diaphragm of mammals have been used in vitro, especially in earliest stages of development and in respect to changes of ACh sensitivity (112, 113). Nerve-muscle cell preparations (organ cultures) in mammals have not been successful; however, such complex mammalian tissue cultures, used first by Harrison (114), have opened important new approaches (3-5, 31, 32) concerning formation of neuromuscular contact, especially morphological and electrophysiological; but much less is known about metabolic aspects of neuromuscular connections. Mammalian (5, 115) and avian (116, 117) "complex" (i.e., spinal cord and muscle tissue) systems have also been used.

Time of differentiation of muscle is another factor affecting the choice of the animal. The muscles of the chick are already differentiated according to speed of contraction into fast and slow muscles at the time of hatching (87), whereas the differentiation is almost absent at birth in the rat (86) or cat (85). It is, of course, advisable to use animals in which the ontogenetic development of certain parameters and the time course after nerve section is well known. From this point of view, the rat is suitable in respect to development of contraction properties (86, 118) and the chicken in respect to that of enzyme patterns (119).

Knowledge of the changes of physiological parameters during denervation and reinnervation is important for interpretation of the metabolic results. These changes follow a definite time sequence, different for different physiological parameters; their correlation in time with metabolic changes in neuron and muscle provide important clues on mechanisms of nerve-muscle interactions. Little is known about the neuronal changes, the cat providing the main data. The failure of the monosynaptic reflex is recorded 72-90 hr after dorsal root section (120, 121); after section of motor neurons this failure is observed for some weeks (122). These experiments show a temporal dissociation between "synaptic" and "trophic" functions of the neuron after nerve section in which a temporary increase of synthesis of ribonucleic acids and proteins is observed (123) at the time of failure of synaptic transmission. The onset of the most important changes in nerve and muscle should be known and experiments designed with respect to these data. The rabbit, rat, and cat provide the main experimental data in this connection. Failure of spontaneous transmitter release [i.e., disappearance of the miniature endplate potentials (m.e.p.p.)] in muscles of the rat occur after about 24 hr (27, 124) and extrajunctional ACh sensitivity from 24-48 hr after nerve section; during this time there is also failure of neuromuscular transmission. The onset and trend of changes in contraction properties differs in fast and slow muscles, e.g., progressive prolongation of contraction properties is observed in the fast muscle and shortening in the slow twitch muscle (68). Fibrillation activity, however, is observed at 29-40 hr (125) and 42 hr (126). Contracture response to ACh and increase of membrane resistance was found until three to four days after nerve section. In the cat and rabbit, neuromuscular transmission fails 32 hr after nerve section; then fibrillation and hypersensitivity (25) appears after 90-120 hrs; nerve conduction, however, about three to four days after nerve section (for review see Ref. 23). In all functional correlations, data on the age of the animal, level of nerve section (distance of nerve section to muscle), and type of muscle must not be neglected.

C. Choice of Nerve and Muscle

The neurochemical implications of denervation, reinnervation, and cross-innervation studies concern especially the problems of neuronal specifity which is revealed by the organization of nerve and muscle cells in motor units and by the possibility of restoring (in reinnervation) or changing (in heteroreinnervation) the properties of muscle cells through neuronal influences. Experimentally, the use of pure "fast" and "slow" muscles, consisting of fast or slow motor units only, should be optimal. This can rarely be achieved. However, exact knowledge of the muscles and neurons is a precondition for successful experiments.

Muscles differ in structural (89, 94), contractile (88, 127-129) metabolic (130, 131), and membrane (88, 127) characteristics. Phasic "twitch" and tonic muscles have been described, especially in submammalian vertebrates and in birds, the related differences concerning especially focal versus multiple innervation, propagated versus nonpropagated response to nerve stimulation, sensitivity to acetylcholine, speed of contraction, and capacity for long-lasting contractures (for review see 127, 128). In mammals, with a few exceptions (129), basically only phasic fast and phasic slow muscles both exhibiting a propagated response, i.e., fast and slow twitch muscles can be distinguished. Two basic properties, speed and long-term maintenance of tension, reflect a preferential adaptation in different muscles to different functional demands. These are accompanied by differences in metabolic response to rate-limiting processes for contraction velocity, primarily or secondarily related to ATP splitting and/or Ca^{2+} mobility (132-134) and/or long-lasting activities dependent on a high level of oxidative energy supply and utilization. In respect to the great variability of structural, biochemical, and physiological properties of muscles and muscle fibers, a strict classification into distinct groups may seem an oversimplification (132-134). It has therefore been questioned whether there is a definite number of types or a whole spectrum of muscle fibers as regards different properties (132). A useful classification is, however, possible in respect to the motor unit organization if one is aware of a tendency of simplification.

Properties of muscle fibers and the neuron innervating them are closely related according to motor neuron organization in the neuromuscular system. These relations are best revealed by comparing the properties of fast and slow motor neurons with fast and slow nerve and muscle cells (Table 1).

Concerning choice of muscle, two sets of observations can no longer be ignored: (a) muscles are to a different degree inhomogeneous, containing only very rarely muscle fibers of the same metabolic type; (b) color (red or white) is not necessarily related to speed of contraction as was earlier suggested by Ranvier (135). Thus, for instance, comparison of "breast muscle" with "leg muscle" (136) or with "predominantly red leg muscle" (137) have a limited value only.

The muscle contains for all practical purposes two or three types of muscle fibers and an inverse relationships between oxidative and glycolytic enzyme activity (92). Fast fibers generally have high glycoltyic (e.g., phosphorylase) and low oxidative (e.g., SDH) enzyme activity. Muscle fibers have a different degree of ATPase activity (91), i.e., high ATPase activity in fast (type II) and low in slow (type I) muscle fibers. As a muscle is composed of many motor units (i.e., the axon innervating about 100 muscle fibers) different degrees of heterogeneity and occurrence of "mixed" muscle types have to be expected, reflecting the influence of innervation on typing of fibers (138).

Speed of contraction of a muscle is not closely related to color and preference of glycolytic or oxidative path ways. It is rather due to a multiple adaptation of muscle to different demands; nevertheless the basic distinction of fast and slow motor units and muscles affords an adequate differentiation. Moreover, it must be remembered, that typing of muscle fibers is not a fixed characteristic. Fiber "types" not only undergo changes during denervation, reinnervation, and cross-innervation (54, 138, 141, 143), but rapid alterations in histochemical distribution occur also as a result of exercise (142) and compensatory hypertrophy (144). This dynamic change, especially in oxidative enzyme activities, makes a nomenclature based on static "fiber type" characteristics doubtful (145).

We have still no definite classification of fibers which would satisfy all requirements. Red muscles, e.g., the laryngeal muscle of the rabbit, may be extremely fast (146) and show extremely high ATPase activity of myosin (147). A differentiation on the basis of energy-supplying enzyme activities is certainly insufficient and may be misleading. It appears that a classification based on motorneuron characteristics, related to speed of contraction, is the most useful one.

The heterogeneity of muscles does not affect or limit this trend in classification. The motor unit is homogeneous, according to selective histochemical response in muscle fibers of one type to stimulation of a single axon and the properties of neuron and innervated muscle cells are closely related (148). Thus the physiological, metabolic, and also structural specializations studied are closely related to the functional roles played essentially by fast and slow motor unit types. Table 1 shows the basic differences. Least studied is this differentiation in nerve cells; it must be remembered that due to multiple adaptation, intermediate units occur (Fig. 8).

1. Choice of Muscle

It must be remembered that in many respects, e.g., speed of contraction, myosin ATPase activity, molecular characteristic of myosin subunits, and also energy-supplying enzymes, fast and slow muscles show differential behavior after denervation (68, 141, 149, 150). For instance, speed of contraction and myosin ATPase activity decrease in the fast, but increase in the slow muscles of rabbit or rat. This factor may explain why change in oxidative metabolism occurs in the fast but not in the slow muscle innervated by a foreign neuron (151).

Thus, it is not possible to state generally that denervation causes slowing of contraction time (152, 153). There is a general decrease of enzyme activities in the denervated muscle (154), but these data are not completely relevant if no time gradient and no exact shift of histochemical enzyme pattern are studied. The importance of the time gradient was realized, e.g., by the finding of a decrease of oxidative enzymes after 60 days (151) and a recovery to normal values after 120 days in the cross-innervated soleus muscle of the cat (153).

Changes in speed of contraction of muscle during denervation, reinnervation, and cross-innervation are very marked indeed; they stress the neural influence and reflect specific electrophysiological and, as yet little studied, metabolic characteristics of related fast and slow motor neurons.

These considerations will affect the choice of nerve and muscle in denervation and reinnervation experiments. The use of muscles composed

TABLE 1

Summary of Differences between Fast and Slow Muscles and Nerves (mainly mammalian)

	Fast	Slow	Ref.
Motor–unit: activity pattern			
Preferential adaptation	speed	maintenance of tension	128
Type of activity	phasic	sustained	
Neuron– excitability	low	high	96, 99, 100, 228
inhibition capacity	marked	smaller	99, 100, 228
frequency discharge	high (in bursts)	low (duration)	96, 99, 100, 228
fatigue resistance	low	high	98, 148, 160
adaptation	rapid	slight	228
Axon– conduction velocity	fast	slow	228
Muscle: Excitability and contraction			
Impulse activity	propagated	phasic–propagated; tonic–non–propagated	88, 127, 129
Ca^{2+} influx	phasic	long–duration	127, 133, 134
Ca^{2+} binding in SR	high	low	127, 133, 134

Muscle: Excitability and contraction

Mechanical activation	at threshold	graded (in tonic)	127
contracture (ACh, K$^+$, caffeine)	0	in tonic; rare in phasic	44, 88, 103, 129
twitch contraction time	short	long	49, 85, 86, 87, 118
relaxation time	short	prolonged	49, 85, 86, 87, 118
max. tetanic tension	higher	lower	127, 98
Myosin ATPase activity	higher	lower	190, 191, 193, 195
at alkaline pH	stable	labile	54
light chains (subunits	different components		192, 247
3-methyl-histidine	high	low	248

Muscle: Energy Supply and Utilization

Glycogen content	high	low	51, 87, 130
Glycolytic enzyme activity	raised	lowered	130, 131
TCC enzyme activity	lower	higher	119, 130, 141
fatty acid metabolism	lower	higher	130

TABLE 1 (Continued)

	Fast	Slow	Ref.
rate of blood flow	lower	higher	212
oxygen consumption	lower	higher	130
color (myoglobin)	> white	> red	130, 135
Biosynthetic Mechanisms			
Neuron– rate of protein synthesis	lower	higher	179, 180
Muscle– rate of protein synthesis	lower	higher	87, 257
Muscle– protein degradation	lower	higher	256
Muscle– rate of developmental differentiation	slower	faster	258
Structural Characteristics			
Neuron size	large	small	100, 161, 228
histochemistry	> glycolytic	> oxydative	161
Axon	larger	smaller	228
Neuro–muscular junction	focal	multiple (tonic) focal (phasic)	88, 89, 134
vesicles in terminals	numerous	less numerous	94
junctional folds	deep–numerous	shallow–sparse	94
synaptic area	larger	smaller	94

Structural Characteristics

Muscle fiber			
diameter	larger	smaller	94
fibrils	regular-separated	irregular	89, 132, 134
sarcopl. reticulum	abundant	less	132, 134
transverse tubular system	regular > triads	irregular dyads	132, 133, 134
Z-line	narrow	wide	94, 134
mitochondrial content	low	high	94, 95
histochemical type	more hetero-geneous	more uniform	
	> type II	> type I (ATPase)	91
	> type A	> type B, C (SDH)	93
	> "white"	> "red"	94
	> phosphorylase	< phosphorylase	92

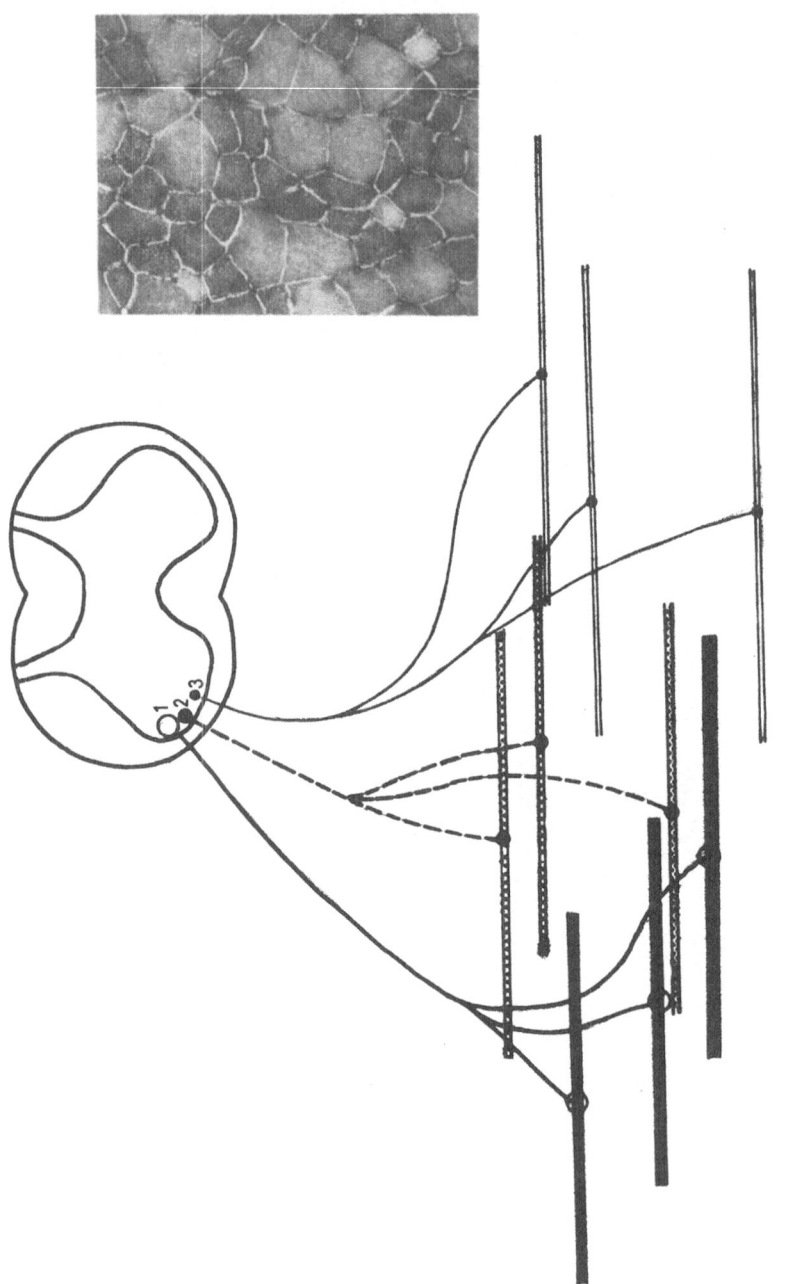

FIG. 8. (a) Scheme of distribution of three motor units causing a three-type muscle fiber pattern in the muscle. (b) Distribution of three types of muscle fibers according to degree of myosin ATPase activity in the extensor digitorum longus muscle of the rat.

only by fast or slow motor units or of muscles with homogeneous units would be ideal. According to contraction properties three types of motor units appear to exist (96-98) with related ultrastructural features (94). Generally a mosaic pattern of three different types of muscle fibers is found in "mixed" muscles. In the soleus muscle of the rat the majority of motor units are slow with a minor proportion — about 10% of intermediate units only. The soleus muscle in the adult guinea pig is homogeneous according to histochemical ATPase fiber type differentiation but contains two types of ATPase activity at birth. This developmental differentiation toward a single type is impaired following neonatal neurectomy (143). The diaphragm or the semitendinosus muscle of the rat or gastrocnemius muscle of the cat has three different fiber types (94). Thus the use of muscles composed only of fast or slow motor units is to be preferred.

2. Choice of Nerves

The use of nerves innervating one muscle, e.g., in the rat the soleus or extensor digitorum longus muscles have advantages. The peroneal nerve innervates practically only fast muscles; cross-union to the tibial nerve will give more reliable results than cross-union of tibial to peroneal nerve, since the former nerve innervates both slow and fast muscles. For reinnervation studies, the nerves should be crushed or sutured near the entry into the muscle to avoid unnecessary damage to blood supply. In cross-union studies, the foreign nerve is cut before entry into the muscle, mobilized for a long stretch of nerve, and then put in apposition to the nerve to be reinnervated higher up. Thus, stretching of the nerve is avoided and a strong union without stress can be performed by silk sutures or, in the case of thin nerves, by sutureless methods, i.e., plasma clot or fibrin union.

D. Biochemical Methods

Biochemical methods are discussed only in respect to denervation and/or reinnervation experiments, primarily considering their relevance to problems of metabolic nerve-muscle cell relationships. In vivo studies by far outnumber in vitro studies; however, it is increasingly realized that in in vivo studies the neuron and muscle cell are exposed to multiple — often poorly understood — regulatory influences. More relevant information will therefore be gained by organ and tissue-culture studies which isolate the cells from the control systems present in the whole animal (see e.g., Refs. 4, 5, 31, 32, 111, 116, 117). Relevant methods are discussed separately for neuron and muscle cell according to specific experimental requirements and conditions of denervation and reinnervation.

1. The Neuron

The choice of neurochemical methods used in denervation and reinnervation studies of muscle should reflect and relate to

(a) the heterogeneity of cell populations in the spinal cord,

(b) the tremendous capacity of peripheral neurons to synthesize proteins and ribonucleic acids which represent an "intensified expression of a continuously ongoing growth process" during regeneration (155), recovery, and maintainence of a cell process in man to 100 cm length. The axoplasmic flow, interaction with the satellite Schwann cells, and the signals for "chromatolysis" ascending from axonal injury to cell body (156) have to be considered;

(c) the sequence of metabolic changes in the cell body occurring after nerve section and regeneration (157, 158),

(d) the metabolic specifity of the motor unit demonstrated in a close correlation between properties of motor neurons and the muscle fibers innervated by the same cells (98, 148, 159, 160, 161),

(e) the neurogenetic implications introduced by the "specifity" of motor neurons made possible by the use of self- and cross-union reinnervation experiments.

(a) Taking into account the marked heterogeneity of the spinal cord cell populations, the preferred methods for studies on isolated neurons are free-hand dissection of neurons from fresh material (162), fixed tissue (163), or frozen sections dried in vacuo (164), or histochemical or histoautoradiographic studies of the spinal cord (161, 165). In studies of rate of breakdown and synthesis of proteins and nucleic acids, biochemical and histoautoradiographical time gradients of incorporation of precursors into RNA and proteins of isolated neurons would be most suitable. They must take into account so far unsolved problems of estimations of specific activity of labeled aminoacids in specific compartments of precursor pools (166), heterogeneity of turnover rates also observed in spinal motor neurons (167), as well as the effect of extraneural (e.g., hormonal) factors on the kinetics of protein synthesis, especially in vivo experiments. The difficulties, of course, increase if fractionation of RNA populations is attempted on a cellular basis. This becomes necessary to study the elaborate neuronal machinery with respect to its directing the specifity and correcting initiation and termination in the translation of an RNA message into a polypeptide product. However, such a discussion is beyond the scope of this article.

(b) The dramatic reaction of a neuron after axon injury, the "chromatolysis," is essentially a marked anabolic reaction providing the increased metabolic activity needed for axon regeneration (8, 156). For instance, neuronal RNA content and rate of protein production may double in response to axonal injury (168). This response has been used as an anatomical marker for mapping the distribution of nerve cell bodies in the spinal cord. Biochemical and histoautoradiographical incorporation studies are our main source of knowledge and have to be done at a cellular level. It must be remembered that the ventral horn cells react differently according to developmental stage, size of the nerve cell, and type of animal (107, 169).

The extremely high level of protein synthesis in neurons is related to the movement of proteins from cell body into the axon. This axoplasmic flow has been indicated, e.g., by the damming experiments of Weiss and Hiscoe (22). Radiographic studies provide further evidence; ^{3}H-leucine has been especially used in such studies (170). Rate measurements of the axoplasmic flow reveal a broad spectrum, from several millimeters per day up to 900 mm/day (23, 171), related to movement of different structures. Fractionation methods have therefore to be used for identification (171). As this rate is affected during regeneration of the nerve (172), knowledge of it will become increasingly important. The difficulties in interpretation concern Schwann cell contribution and extracellular rate of labeling. In vitro studies of nerve-muscle preparations of snail (47) and frog (173), using silicone-grease barriers to prevent leaks between the compartments to which radioactive metabolites are added, therefore provide many advantages.

(c) Reinnervation and cross-innervation studies may be performed at different times after nerve section in order to study the sequence of metabolic changes occurring after nerve section: knowledge of the time gradient is therefore necessary. The anabolic retrograde reaction after nerve section in the nerve cell occurs within 1–3 days (8), good indicators being uptake of precursors into proteins and RNA, increase in number of mitochondria, and acid phosphatase activity (174). The pattern of the metabolic changes is complex: anaerobic glycolysis is apparently increased (175), and aerobic glycolysis decreased (176). At longer times, two to three months after nerve section, there is a decrease of incorporation of amino acids into proteins and loss of nucleoproteins, but with regeneration of the nerve recovery of normal cell function is achieved about 80 days after a nerve crush (177), at least according to the absorption spectrum of the Nissl substance. A gradient of detailed incorporation studies would be necessary for full understanding of this process. The intensity of the "retrograde" reaction differs according to cell type, age, type of lesion, distance of section from nerve cell, delay of nerve suture, and other factors (58). Cell death occurs at different rates in different cells. Some central neurons degenerate very late, e.g., in the optic tract, others very early, e.g., in the olfactory receptor neurons. Very different responses are also seen in transsynaptic retrograde degeneration. Cell death is more frequent and increased rate of incorporation of precursors much more pronounced in neurons of young animals (178). This observation may be relevant to the nature of the signal for chromatolysis, which probably concerns loss of presynaptically derived trophic substances or loss of a neuronal repressor substance (156). For the proper design of experiments, it is important that the proportion of neurons undergoing cell death is rather high, but it is similar after section or after crushing the facial or hypoglossal nerve. Section or crushing of a long peripheral nerve near the entry into muscle result in least intense retrograde changes. This is apparently due to the relatively smaller loss of proteins of the neuron.

(d) Neurons are characterized by a high degree of specifity of structure and metabolism. Denervation and reinnervation experiments have to take into consideration (1) the characteristic features of motor neurons (especially the fast and slow α motor neurons, but also interneurons including Renshaw and γ fusi motor neurons; the first two groups are in the center of our interest) and (2) the homogeneity of motor units revealing conformity of physiological and metabolic properties of the motor neuron and the muscle cells it innervates. No doubt, considerations of these relations will be of great importance in evaluation of reinnervation and cross-innervation experiments. The physiological properties of the two groups of α motor neurons differ according to axonal conduction velocity, neuronal input resistance, post-spike hyperpolarization, fatigue resistance, and differential activation into reflex systems (160). Less is known about the metabolic differentiation of these two types of motor neurons. Histochemical and incorporation studies give some but, so far, only preliminary information. Neuronal size is an important factor. Neurons larger than $30\,\mu$, presumably fast α motor neurons, show strong phosphorylase and weak succinic dehydrogenase activity, whereas neurons smaller than 30μ, presumably slow α motor neurons, reveal a reciprocal behavior (161). There is also a correlation between motor neuron size and rate of protein synthesis studied histoautoradiographically (179, 180).

(e) The close correlation between a number of properties intrinsic to the two groups of α motor neurons and the muscle fibers innervated by these cells are of great neurochemical interest. The exact correlation of neuron and related muscle fiber characteristics will need further, more exact histochemical study of RNA and protein metabolism in the different motor units. The main functional-histochemical correlations in muscle fibers depend mostly on studies of ATPase and oxidative enzyme activities. For instance, the population of gastrocnemius motor units could be divided into two or three types, i.e., two types of fast contracting units, — type "FF" showing low fatigue resistance, high ATPase, and low oxidative enzyme activity; a second (type "FR") with high ATPase and high oxidative activity; and a third motor unit (type "S") with slow contraction time, great fatigue resistance, low ATPase, and high oxidative enzyme activity, re-sembling characteristics of the slow soleus muscle fibers (98). Muscle fibers of the "FF" units generally depend primarily on anaerobic glycolysis and are predominantly "white" muscles; the "S" type units depend on aerobic pathways for energy. Thus motor units may react to different me-tabolic demands, revealing a "multiple adaptation." A more complex pat-tern of motor units may be expected (128), but it appears to be rather a modification only of the basic pattern of fast and slow motor units. The choice of muscle will have to refer to these motor unit characteristics such as neuron size, histochemical properties, and rate of incorporation. Many more data are necessary for a complete analysis of a motor unit in neurochemical studies.

2. Muscle

A very wide spectrum of metabolic changes in muscle following dener-vation and reinnervation has been described [for review see (8, 58, 181)] comprising changes in water content, electrolytes, proteins and nucleic acids, metabolites, and enzymes related to mechano-chemical mechanisms and energy-supply and utilization, etc. The significance and comparability of these changes is increasingly questioned, especially in respect to (a) the lack of uniformity of intervals in different studies which make it diffi-cult to establish precise time relations, (b) different time courses of changes in different animal species and muscles, (c) the inhomogeneity of muscle according to type of fiber distribution, and (d) the compartmentali-zation of the different constitutents of the muscle fibers. The present paper can, of course, give no full evaluation of the different methods, but some of the inherent difficulties and the prospective possibilities of those meth-ods are discussed, which, in time, will have neurochemical implications.

Atrophy after denervation and recovery with reinnervation reveal es-pecially changes in control of breakdown and synthesis of proteins. Whole muscle changes have limited importance, as a differential response to de-nervation occurs in soluble and contractile proteins (182, 183). The total protein content is with 60-70% of the tissue largely localized in one struc-ture, the myofibril. The predominance of myofibrils, which upon homoge-nization tend to trap organelles preventing their differential centrifugation, has led to difficulties of muscle tissue fractionation (for methods see 184). The difficulties in RNA separation have been essentially reduced by in-creasing the ionic strength of the homogenization medium (185). Increase of the rate of degradation of proteins appears to be established by findings of increased activity of proteolytic enzymes and peptidase activity of sarco-plasmic proteins (87), as shown by disappearance of radioactivity from muscle proteins labeled with ^{3}H-leucine prior to section of the nerve, especially from myofibrillar proteins (186). Data on the overall rate of synthesis are more difficult to interpret. The main features of denervation atrophy appear to be an initial increase of RNA-controlled synthesis of proteins (187), apparently due to increased substrate supply. The ratio of the free amino acid pool to specific activity of structurally defined muscle proteins and the alterations in transport of amino acids during denervation (188) should be known. Such data may explain controversial findings reporting both increases (183 and others) and decreases (186) of synthetic activity in denervated muscle. Moreover, protein synthesis proceeds in a different manner for different amino acids (189) and different precursors RNA (187). However, interpretations are made easier due to the findings that amino acids are incorporated into proteins directly from the extracellular pool (189). Myosin from fast and slow muscle shows several differences in molecular structure (190-192). In view of the biosynthesis of myosin mediated by different types of neurons, it is essential to have information on neural effects on the emergence of the contractile proteins, on their capacity to make ATP rapidly available for contraction, and on the calcium transport system (193). Ultrastructural research has brought many new data on synthesis and assembly of myofibrils in embryonic muscle (194). However, biochemical studies at a molecular level are few for technical reasons. The structure of myosin and its behavior is apparently changeable in response to requirements and activity pattern of the cell. Detailed studies, e.g., on low-molecular-weight components, enzymic properties of the myosin molecules, myosin isoenzymes, susceptibility to tryptic digestion, methylhistidine content, etc., did therefore prove important, especially when neuronal regulations were detected (55, 195, 196).

The significance of studies concerned with changes of energy metabolism is reduced, as the changes are mostly not related to number, diameter of muscle fibers, and size and composition of extracellular space. A comparative analysis of enzyme levels and activity ratios of key enzymes in energy-supplying metabolism has resulted in classification of different "metabolic" types of vertebrate muscles. Based upon the highly discriminative activity ratios of enzymes representing energy supply systems, "system relations," (e.g., triosephosphate dehydrogenase/3-hydroxyacyl-CoA dehydrogenase, triosephosphate dehydrogenase/citrate synthase, or glycogen phosphorylase/hexokinase), a high specifity for muscle types was demonstrated (131, 197). (For methods of measurements of different enzyme activities see Ref. 198.) These activities should optimally be referred to single muscle fibers. The number of muscle fibers should be known, but this has rarely been achieved. The data should at least be referred to, both total muscle weight and weight per unit.

For instance, after castration of male rats resulting in atrophy of the hormone-sensitive levator ani muscle, atrophy, but not decrease of number of muscle fibers may be observed; concomitantly, choline-acetylase activity decreases by 41% when referred to the whole muscle but increases by 112% when referred to unit weight (199). Histochemical analysis does refer to single muscle fibers; however, it provides practically only qualitative data. These considerations apply also to reinnervation experiments during which levels of metabolites, rate of degradation, and synthesis of proteins return to normal values (8).

As has been pointed out, denervation and reinnervation experiments have neurogenetic aspects, and studies on fixation and/or transfer of genetic information will become increasingly important. Such studies should concern data on changes in (a) DNA structure, nucleotide base-composition, or DNA polymerase activity, (b) transcription, and (c) translation mechanism. The fractionation of RNA is of course a precondition for such studies; however, little work has been done in muscle. Studies orientated in this way should uncover genetic mechanisms in the regulation of protein synthesis of the muscle cell. The difficulties inherent to studies of muscle tissue are due to the relatively low amount of RNA in muscle (200), a smaller amount of mRNA, greater technical problems in isolation of mRNA, and contamination with myofibrils. Studies on polyribosomes which have more than 100 monomers in muscle are especially important. The synthesis of myosin on polyribosomes has indeed been described in myogenesis (201). These polysomes may be isolated on the basis of sedimentation in the sucrose-gradient. The difficulties encountered in muscle with cellular system are partly overcome by RNA synthesis in cell-free systems (200) which allow standard addition of ATP and transfer enzymes to the different systems. Important data are obtained by studies of myosin biosynthesis where optimal conditions for incorporation of labeled amino acids can be achieved in a medium of high ionic strength (185).

Information on rapidly labeled RNA, especially the time course of labeling in muscle, would be very important in respect to regulations on a molecular level. The time course of labeling has been investigated more in brain (202) and other tissues than in muscle (203, 204). For such studies, fractionation of muscle RNA by polyacrylamide gel electrophoresis (205) is to be preferred to sucrose density-gradient centrifugation methods for higher resolution and sensitivity. In vitro studies have considerable advantages in ensuring stable precursor supply during the time of incubation, and in enabling a study of the temporal pattern of RNA labeling. The presence of polydisperse, rapidly labeled RNA has been observed in different tissues (206), but only lately in muscle (204). In short-term incubation experiments, such an RNA, part of it with a sedimentation coefficient suggesting newly formed mRNA and with a nuclear base composition approaching that of DNA similar to RNA ("template" RNA) was found (204).

The promising experiments on the effect of hormones (especially growth hormone, insulin, and testosterone) on RNA metabolism suggest a

primary action of hormones through stimulation of specific genetic transcription with the consequent production of specific mRNA. It may be expected, that similar trends will increasingly be utilized in studies of muscle (207). However, in muscle not even the use of incorporation into cell-
free systems and polyacrylamide-gel electrophoresis has shown qualitative
changes in pattern of protein synthesis after testosterone application (208).
Evidently, the study of rapidly turning over mRNA in muscle should give
promising results. The disadvantage of studies of neurally induced
changes is, of course, the fact that the neurotrophic agent has not yet been
chemically identified. Indirect approaches, i.e., denervation and reinnervation experiments have, however, not been fully exploited.

E. Side Effects in Denervation and Reinnervation Studies

Side effects, not directly related to changes in nervous regulation include: (a) autoregulative and hormonal mechanisms which interact with
nervous influences, (b) vasomotor mechanisms, (c) immobilization effects
affecting the neuromuscular and other systems of the organism, (d) peripheral mechanisms operating, for instance, by changing length of the muscle
fibers, (e) compensatory mechanisms. A differentiation between these
and the direct nervous mechanisms is often difficult but necessary to establish specifity of denervation changes. Only a few examples may be
quoted.

(a) An autoregulative mechanism is, for instance, the resynthesis of
metabolites after different stimuli (e.g., resynthesis of glycogen after its
breakdown following muscle stimulation). This will, of course, also occur
in a denervated muscle; however, overshoot reactions (resynthesis above
starting levels under comparable conditions) is observed only when innervation is intact (8). The effect of hormones is demonstrated either directly,
i.e., in the cause of special target muscles with high sensitivity to the
hormone, or generally, in all muscles during early stages of development,
e.g., due to the anabolic action of growth hormone. Androgens have a
marked effect on the rate of protein synthesis and contraction properties
of the highly sensitive levator ani muscle of male rats (209). This muscle
undergoes complete postnatal involution in the female rat; however, its
differentiation and maintenance can be ensured even in the female rat after
perinatal application of androgens (211), and this effect can be demonstrated even when denervation is performed at birth (210). In this case the
hormone temporarily has a relatively nerve-independent trophic effect.
The latter is lost at subsequent stages of development; on the other hand,
weight loss in denervated skeletal muscles may almost be absent in early
stages of development. Our interpretation of denervation experiments in
early stages of development therefore has to consider the special hormonal
environment at this period.

(b) Blood flow increases in denervated muscles (212) and affects the rate of substrate supply. This factor operates differently in different muscles. Resting oxygen consumption as measured in arteriovenous differences is much higher in fast than slow muscles and this could be explained by a much higher blood flow in slow muscles. Differences in blood flow and the related substrate supply explains why changes in rate of protein synthesis have a different course in vivo and in vitro, the latter being unaffected by the increase of blood flow (213).

(c) Long-term immobilization effects due to motor inactivity connected with denervation of limbs results in complex changes affecting the neuromuscular and other organ systems. Reduction of motor activity leads to muscle and probably neuronal atrophy. Denervation may in some respect produce changes resembling "restraint stress," with increased liberation of corticotrophic hormones and related catabolic effects. Restraint stress decreases protein synthesis in brain cortex slices (214) and in spinal cord neurons (215). Effects of immobilization may operate also in the development of muscle atrophy in nontarget muscles after castration,. Castration results in marked reduction of motor activity. The much more marked atrophy in the target muscles (e.g., the levator ani of the rat or the temporalis muscle of the guinea pig) is, however, related to the high androgen dependence of these muscles (216). Immobilization, just as denervation, may also result at certain periods in an increase of blood flow in muscle. More general effects are loss of calcium from bones, development of contractures, and the disturbance of hematopoesis. This "post-denervation anemia" (217) is a temporary event, the hypochromic anemia being probably a result of adaptation to reduced motor activity (181). Changes in the internal environment have to be considered in denervation experiments, especially when testing the effect of hormones or other agents in respect to change of substrate supply and/or decreased caloric intake during immobilization. For instance the somatotrophic hormone reduces rate of denervation atrophy even in adult animals during realimentation following starvation (181). No full account of these factors can be given but they have to be considered in denervation experiments in vivo.

(d) Stretching of a muscle in vivo and in vitro increases the rate of protein synthesis (144) and accordingly a different degree of stretch affects the rate of denervation atrophy. Thus, compensatory hypertrophy after elimination of synergists can be observed even in a denervated muscle (144, 218). The degree of overlap of filaments varies, e.g., in a muscle tenotomized or stretched, and this factor also affects both ATPase activity and contractile behavior in a denervated muscle (144). This mechanism apparently operates in the denervation hypertrophy of some muscles also (187, 213, 219–223), i.e., the diaphragm and the latissimus dorsi anterior of the chicken. It might operate also in other muscles undergoing stretch in long-lasting denervation.

(e) "Compensatory hypertrophy" of the control ("contralateral") muscle may be a factor in quantitative studies of changes in a denervated muscle. This factor has not been studied in detail, but it is advisable to use also unoperated groups of animals to increase the control value of the "contralateral" muscle. Changes in enzyme activities (145) and contractile behavior in the "control" muscle of the contralateral side may be considerable indeed. Sprouting of nerve fibers from neighboring nerves can also be interpreted as a compensatory phenomen. Such nerves may accomplish heteroinnervation if the muscle is denervated and may under certain conditions produce trophic effects even in innervated muscles without restoring synaptic neuromuscular connections (224). Ligature of the surrounding nerves is the only method of avoiding such phenomena. However, a few vasomotor nerve fibers will be found near vessels in all denervation experiments. Such fibers originate from very different ganglia and reach the muscle in addition to sympathetic fibers accompanying the muscle nerve. Their effect can, however, be neglected. Neither sensory (225) nor sympathetic neurons appear to affect denervation (226), the trophic influence of nerve on muscle being essentially a property of axons of motor neurons (227).

III. DISCUSSION AND INTERPRETATION OF RESULTS

A. Contributions and Limitations

Why should one perform denervation and reinnervation experiments from a neurochemical point of view? The study of disturbances of neuronal and muscle cell metabolism after nerve section and its recovery with reinnervation appears, indeed, to provide adequate models for the study of interactions between nerve cells and nerve and muscle cells, i.e., neurocommunication. However, they allow differentiation of two basic groups of communication. The metabolic energy for the first type of communication is related to transmission of nerve impulses and is primarily concerned with the operation of ionic pumps involved in maintaining the ionic concentrations of nerve cells and the electrical potentials across the surfaces of membranes and with the synthesis both of specific transmitters and their receptors. The metabolic energy for the second type is concerned rather with synthesis, transport, and possibly transfer at the neuromuscular junction (32, 47) of "trophic" agents which maintain structural integrity of the cells and "initiate" or control molecular modification in other cells (13). These experiments also provide information on long-term effects of use and disuse on nerve and muscle cell and on the degree of specifity and plasticity of neurons and muscle cells.

Denervation and reinnervation experiments contribute to the problem of the "trophic functions of the neuron," i.e., nonimpulse function. Their limitations are due to the fact that denervation and reinnervation are "elimination" and "replacement," experiments (13), respectively, working – in contrast to studies on hormonal regulation – without knowledge of chemcally defined agents. Moreover, the differentiation of intercellular regulations or neurocommunications into shot-term events (related to transmission of nerve impulse activity), synaptic facilatory and inhibitory reactions explained basically by membrane changes (228) and long-term events (i. e., "trophic" regulations maintaining and restoring structure and functional capacity of cells relatively independent of nerve impulse), is still rather schematic (8). Both regulations are normally interrelated. The definition of adequate and convincing indicators of neurotrophic functions thus presents difficulties which are discussed.

(a) It has become clear from several distinct experiments that the neuron operates not only in mediator synthesis and transmission of nerve impulses, but also affects growth, differentiation, and maintenance of tissues independently of this activity. These include: experiments on the importance of nerves for regeneration of amputated newt limbs (29), structural integrity of receptor organs (1), regeneration of minced muscle tissue (229), and denervation and reinnervation experiments of cross-striated muscle quoted above (8, 9, 13). A basic role of impulse-related transmitter release for trophic functions is improbable; however, the question of the trophic function of spontaneous transmitter release is more difficult to answer. Experiments on regulation of sensitivity to ACh and synthesis of cholinesterase in vivo (12, 13) and in vitro (31, 32) suggest that the "trophic" intercellular communication process is not necessarily dependent on synaptic transmission. The existence of trophic effects, mediated by diffusible chemical substances, produced by the nerves, and acting without synaptic contact, is suggested (32). However, the mechanism of "transfer" of such substances at the neuromuscular junction (47, 48) is not yet proved.

(b) A clear definition of the trophic function of the neuron cannot be achieved without clarifying the effect of use and disuse on nerve and muscle cell and the mechanism mediating such effects, either by transmitter or neurotrophic activity. If no agent other than the neurotransmitter is responsible, or if the denervation changes can be reproduced by disuse alone, the postulation of a specific trophic function of the neuron would indeed lose its impact (15). In this respect the significance of denervation and reinnervation experiments is not yet exhausted. Disuse (in the sense of absence of impulse-directed transmitter release) does not reproduce denervation changes in the muscle (8) and it is improbable that loss of spontaneous transmitter release might be the decisive "neurotrophic agent." The increased sensitivity to ACh over the whole muscle membrane (46), the marked disturbance of membrane stability leading to fibrillation (45), and some intracellular changes, e.g., the increase of DNA in muscle (59),

appear to be specific for denervation atrophy. They cannot be reproduced
by "disuse" (8, 9, 12). Moreover, attempts to substitute ACh for "trophic"
actions in the sense of a "replacement strategy" (13), which would explain
the neural regulation of ACh sensitivity and/or cholinesterase activity in
muscle, have failed. Keeping a denervated muscle in vitro in ACh solutions
does not reduce the chemoreceptive zone, which is enlarged after nerve
section (230); infusion of ACh in vivo to rats poisoned with diisopropyl-
fluorophosphate has no effect on resynthesis of cholinesterase in muscle
(231), nor does the addition of an analog of ACh to monolayer cultures of
embryonic muscle cells (232).

The finding of miniature endplate potentials (m.e.p.p.) in completely
denervated muscle (233), the discrepancies between the m.e.p.p. produced
by spontaneous transmitter release and chemosensitivity (10), the rapid
decrease of ChE activity with denervation but not with disuse (9), and many
other findings suggest that spontaneous transmitter release does not me-
diate the decisive trophic effect of the neuron (13, 39) and that denervation
atrophy cannot be identified with "disuse" atrophy. A metabolic disturbance
due to lack of neurotrophic agents released from the nerve must at least
participate. Pharmacological denervation, e.g., with botulin toxin does
reproduce specific denervation changes (63); however, the toxin may block
the release of neurotrophic agents other than ACh, together with elimination of cholinergic transmission (13). It must, however, be admitted, that
the evidence for neurotrophic agents is, so far, only an indirect one.

In studies of the effect of use and disuse two conditions should be kept
in mind, i.e., a careful definition of the model of disuse and critical choice
of the criteria studied. For instance, tenotomy of a muscle is not an adequate model as it leads to increase of excitability of motorneurons and to
marked shortening of the muscles. Denervation and reinnervation provide
the neurochemist with important models of "use" and "disuse" affecting
neuronal metabolism. Neuronal atrophy after nerve section following a
phase of increased RNA and protein metabolism, the ultimate cell death
in some and survival in other neurons, the recovery of metabolic characteristics during the process of regeneration, all of these findings will give
increasingly more important neurochemical data on correlation between
activity and metabolism in neurons as well as muscle cells.

Muscle atrophy is observed in very different conditions and, by itself,
is not a useful criterion of effect or loss of neurotrophic control. For
instance castration, i.e., elimination of testosterone secretion, leads to
marked muscle atrophy; however, synthesis of ACh in the nerve endings
is affected but little (199). Speed of contraction is undoubtedly neurally
controlled (49) by frequency of firing or total impulse activity (234, 235)
but probably cannot serve as an adequate indicator of neurotrophic influences, as changes in speed of contraction and related ATPase activity

occur relatively late and as differential responses to denervation can be observed in different muscles (68).

(c) The broad spectrum of metabolic properties of muscle and muscle fibers became interesting to the neurochemist when it was shown that, e.g., the enzyme pattern of energy metabolism and also the enzymic properties of the myosin molecule vary according to nervous activity pattern, that enzymic composition of muscle can be modified by an adaptive response to activity (193), and that metabolic types of muscle are related to the metabolism of different neurons. Even in the case of myosin it appeared probable that a process of nerve-muscle cell adaptation might select the functionally adequate type.

It is understandable that in this connection problems of specifity and plasticity of neuron and muscle cell have been approached by reinnervation studies. These questions of great neurochemical interest can only be answered at a molecular level "tracing in detail the pathway from the action of the gene to the final expression of its phenotype" (236).

B. Neurogenetic Aspects

Neurogenetic considerations did arise relatively early in studies on the trophic function of the nerve cell in relation to the loss of metabolic differentiation during denervation and the return of characteristic speed of contraction and metabolic properties during reinnervation in muscles and muscle fibers of different function (39, 44, 49-54, 141, 151, 153, 196, 237-239) (see Table 2). This "specifying" influence of nerve is well known in ontogenetic studies where a high selectivity of neuronal pathways and end organs develop. A previous concept of "genetic neurology" (240) implied that "... different neuron types are fundamentally different in their composition; that a similar critical diversity exists among pathways and terminal organs, and that the neuron must possess means to identify or recognize, as it were, the appropriately matching specifity of its predestined pathway or terminal." It is only now, with the development of molecular biology, that these generalizations can be formulated in terms of a genetic regulation of differentiation involving the synthesis of tissue-specific proteins. Basic problems of origin, specifity, and modifiability of intercellular relationships have especially been studied in the central nervous system and are still to be settled (241, 242). Especially cross-innervation studies of muscle have opened new neurochemical approaches to the problem of neuronal and muscle specifity, suggesting that the nerve influences gene expression in the muscle cell. The evidence for a "successful" cross-innervation experiment will have to rest on exclusion of experimental artifacts, some of them have been mentioned before, and on demonstration of genetic mechanisms in nerve-induced differentiation and transformation of muscle metabolism.

TABLE 2

Effects of Heteroinnervation (changes produced by cross union
or implantation of foreign nerves into muscles)

Systems affected	Refs.
	General Ref. 13, 39, 55, 94 138, 195, 239
Speed of contraction	49, 50, 151, 153, 237, 238
Contracture response	44, 103
Myosin ATPase activity	50, 196, 239
(histochemical)	53, 54, 139, 140
subunits	55
glycogen content	51
K^+ content	51
Glycolytic and oxidative enzyme activities	52, 151, 153
histochemically	141
Proteolytic activity	39
Lactate dehydrogenase isozymes	195
Endplate structure	244

Before discussing different experiments from these points of view,
we must consider the general assumptions, that all the cells of a given
organism contain the same genetic information and that the phenotypic differences between various cell types are based on the selective expression
of their genetic information (243). Whatever the mechanism and site of
genetic control, the result is the synthesis of a tissue-specific protein.
Moreover, there is a developmental change in genetic regulation of differentiation. This is, for instance, indicated by the demonstration of trophic
effects of sensory ganglia on muscle fibers, e.g., on ChE activity on organ
cultures of muscle (32), or successful cross-union in young chicken (237),
phenomena apparently lost in later stages of development (244).

A direct genetic interpretation of cross-innervation studies is justified
only if synthesis of a specific protein can be demonstrated which originally
did not exist in the muscle. From this point of view some of these experiments are now discussed.

1. Changes in Enzyme Pattern of Energy-Supplying Metabolism

Changes in enzyme activities and even in "discriminative enzyme ratios (52) have to be cautiously interpreted. In most cases it is not certain whether the change in enzyme activities is due to change in amount of newly synthesized enzyme or to change in secondary regulatory control. Many nonneural factors affect glycolytic and oxidative enzyme activities, such as hormones, age, and functional activity. A brief period of activity, for instance, markedly increases hexokinase activity (245) and it is thus unlikely that de novo protein synthesis is involved. Moreover, so far only quantitative differences in energy-supplying enzyme systems have been described in different muscles. Lactic dehydrogenase occurs in different isozyme forms with kinetic preferences favoring reduction of pyruvate or oxidation of lactate, respectively; opposing extremes of M type (reductive) and H type (oxidative) can be separated by electrophoresis. Marked differences do exist in this respect in fast (with prevailing M type) and slow (with prevailing H type) muscles and conversion after cross-union is observed. However, the changes in respect to isozyme pattern, though marked, appear to be of a quantitative nature only (195). Quantitative studies, histochemical (246) and biochemical, on single muscle fibers will become more important.

2. Changes of Specific Structure and Function of Myosin

The main evidence for genetic interpretation comes from this type of experiments. Although differences in molecular weight of myosin itself between slow and fast muscles have not been found, such differences do exist in respect to closely associated proteins and subunits (247) which are electrophoretically distinct in fast and slow muscles. Two bands (A and C) are found in fast-muscle myosin but are absent from slow-muscle myosin, the latter having two bands (r and s) lacking in fast-muscle myosin. After cross-innervation, the myosin acquires associated proteins characteristic of the muscle normally innervated by the foreign nerve (55). A qualitative new type of protein is thus synthesized due to the influence of the foreign neuron. The assumption that this is directed by gene action appears to be justified in this case (55). Further evidence seems to come from studies of ATPase activity. After cross-union, marked changes in ATPase activity of fast and slow muscles innervated by the "foreign" nerve are observed (195, 196). There are also changes in some functional properties of the enzyme (pH profile curve and ATP-induced dinitrophenylation). More specific differences of the enzyme and a change with cross-innervation are revealed histochemically by a differential pH lability. ATPase of α fibers is acid-labile, that of β fibers base-labile (54). Slow muscles of the cat or guinea pig contain only β fibers. As α fibers appear after cross-innervation and as the qualitative differences in pH lability may reflect differences

in protein composition of the enzyme, these experiments seem to strenghten
the evidence for neurally induced regulation of gene expression is muscle
(55). Other approaches concerning the existence of 3-methyl histidine in
myosin of fast muscles only (248), or ATP-induced change in dinitrophenylation (249), will need further confirmation. The detection of qualitative
differences is, of course, dependent on the resolution capacity of the method; however, only the change in qualitative differences of the proteins or
the acquisition of new functional properties will provide conclusive evidence for neurally induced genetic control (Fig. 9).

3. Changes in Synthesis of RNA and Proteins

The high degree of specifity of muscle fibers, i.e., multinucleated
muscle cells containing specialized, contractile proteins, appears to be
under neural control. The nature of this specifity must be related to molecular differences and the question is, how they relate to differences in
genetic activity (250). Specifity of differentiation is lost during denervation, recovered with reinnervation, and changed after cross-union of
nerves; this suggests the operation of genetic mechanisms in the regulation of protein synthesis of the muscle cell. It may be useful to discuss
generally the lines of research necessary for such studies. Two considerations must be kept in mind: (a) The genome is equally represented in
all cells and the diversity of cell phenotypes must derive from the fact
that each cell expresses only a limited amount of its full genetic potential
(251) whatever mechanisms [e.g., de-repression of a normally expressed
genome (252), post-transcriptional repression (251), or negative control
by diffusible regulator substances (243)] affect the selective expression
of genetic information. (b) Synthesis of specific proteins will occur also
in tissue cultures and embryonic muscle without nervous control (194).
However, both tissue culture and in vivo studies show that de-differentiation sets in if neural connections are not formed or maintained for longer
periods [for tissue cultures (117), for development of receptor organs (1),
and for muscle regeneration (253)]. These data stress the point that a
differentiation of genetic information residing both in muscle cell itself as
well as in the neuron is necessary. In both cases the information contained
in DNA is transmitted to mRNA molecules which, in turn, serve as the
template upon which the polypeptide chains are synthesized.

In "de-differentiation" of denervation it will be necessary to look for
the process of replication, transcription, and translation which may be
interfered with and may lead to modifications of DNA and RNA, resulting
in the loss of normal biological activity of the proteins. The increase in
DNA, proteolytic enzymes, and choline receptors in denervated muscle
may indicate the operation of a de-repression of a normally repressed
genome supplied by the neuron (8, 59, 113, 254, 255). In this respect it is of

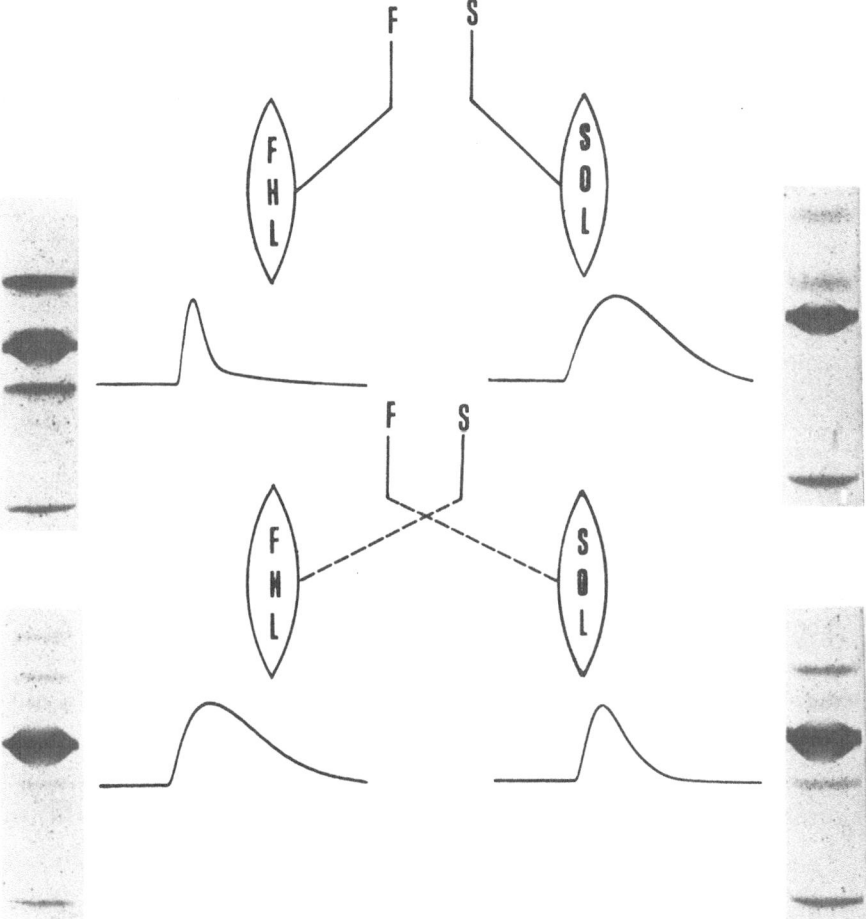

FIG. 9. Schematic representation of cross-union experiment of fast
(F) and slow (S) nerves innervating the slow (SOL) soleus and fast flexor
hallucis longus (FHL) muscle of cats. The records show the resulting
changes in contraction velocity and twitch duration (reprinted from Ref. 290)
and the resulting changes in closely associated low-molecular-weight proteins of myosin, visualized by disk gel electrophoresis. The FHL has four
bands, the SOL two. After cross-union of nerves, the cross-reinnervated
FHL muscle acquires the protein bands characteristic of the slow (SOL)
muscle (left) the cross-innervated SOL muscle those of the fast (FHL)
muscle (reprinted from Ref. 55, Fig. 1).

interest, that some membrane denervation changes, e.g., appearance of
extrajunctional cholinergic receptors in vitro (113) and in vivo (255) and
also the fall of resting membrane potential (255) can be prevented by
actinomycin D, generally considered to block DNA-dependent RNA syn-

thesis. In the "redifferentiation" of a reinnervated muscle genes activated
by a specific "inducer" might make mRNA synthesized in the neuron newly
available to the muscle cell. Concrete experimental data are, of course,
necessary to test these hypothetical concepts.

No direct information on changes of genetic information can be expected
from studies of rate of synthesis and breakdown of proteins and of turnover
rates of proteins in denervated muscles, although this information helps
to differentiate muscle types, e.g., according to higher rate of breakdown
of proteins (256) and synthesis of proteins in slow muscle in vivo and in
vitro (87, 257, 258). Relevant information may be gained by changes in
DNA amount increasing after denervation (59), enhanced incorporation of
^{3}H thymidine into DNA (259), and DNA polymerse activity (187). Studies
on RNA metabolism will become relevant if qualitative changes in the populations of RNA, especially the fast turning over mRNA, base composition,
or labeling pattern of RNA and of proteins, will be detected. Some data in
this line were obtained in studies of changes in RNA and protein synthesis
during induced cardiac hypertrophy (260) but are almost missing in denervation or reinnervation studies. Thus the main evidence for the specifying
of neural control using genetic mechanisms so far still rests on studies of
nerve-induced changes of the myosin molecule only (see Table 2).

C. Open Questions for Neurochemistry

The following open questions and experimental tasks may be formulated.

1. Neurochemical differentiation of the spinal motor neuron types is
necessary in respect to a more precise definition of motor units and of
the metabolic relations of nerve and muscle cell. Single-cell studies appear to be a precondition of such aims.

2. The nature of the signal that ascends from the axon to cell body
and initiates the neuronal reaction after nerve section is still unknown. A
hypothesis that the neuron produces a substance which represses neuronal
RNA production and loses some of this repressor after axon injury has
been put forward (156). Such intracellular, nerve-impulse-independent
signals are apparently operating both ways in maintenance of neuron and
muscle cells and are probably related to the bidirectional character of
axoplasmic transport (23). It is of interest that protein synthesis is 34%
higher in the motor neuron after section close to the neuron than after section distal from the neuron. Loss of proteins may therefore be an important retrograde signal for the neuronal maintenance reaction (261). The
rate with which metabolic and structural disturbances appear after nerve
section depends on the length of the sectioned nerve stump. This suggests
that proximo-distal transport of cellular constituents may act in signaliz-
ing the breakdown of "double dependence" (34) between neuron and muscle
cell. The faster increase of proteolytic activity in a denervated muscle

with a short nerve stump suggests that agents inhibiting degradation of proteins are convected along the axon. Which agents are these and how do they act (262)?

3. The differentiation of impulse-directed and spontaneous transmitter release, the characteristics of spontaneous transmitter release in different muscle fibers, and their relation to sensitivity to ACh and synthesis of cholinesterase will remain an important task. In cross-unions, neurons display great specifity in the induction of characteristic endplate structures (244). New methods are now available for the identification of macromolecules showing capacity for selective recognition of cholinergic agents (263). Genetic mechanisms, as suggested by the cross-union experiments, may be involved.

4. The effects of use and disuse have also been newly approached by cross-union studies (49) and need more neurochemical studies on specifity of neurotrophic influences.

It is improbable that only impulse activity is responsible for the trophic effects mediated by the nerve. Metabolic and functional characteristics of muscle are regulated by many different mechanisms, and it is not likely that the mechanisms operating in the regulation of speed are identical with the mechanisms mediating trophic effects in muscle. A multiple regulation of muscle metabolism and function, in which autoregulative, hormonal, and neurotrophic influences are operating must be assumed. The differentiation of these various mechanisms is still an open question. It is, however, possible that the depolarization of the muscle membrane and changes in membrane permeability occurring during impulse activity may facilitate the effect of other neurotrophic agents apart from quantal release of ACh. The question of how the two mechanisms are correlated and of how the specific effect of the neurotrophic agent and of spontaneous microsecretion of ACh on muscle metabolism interact remain open (262).

5. Denervation, reinnervation, and cross-innervation studies essentially have neurogenetic implications related to the preferential synthesis of specific proteins. In relation to de- and redifferentiation questions and methods of molecular biology of cell differentiation are increasingly introduced.

The presence of redundant genes in eukaryotic cells, the fact that most of the DNA is in a repressed state, i.e., not functional in transcription, the specific fluctuations in cell activities related to different rates and time sequences in synthesis of various RNA species (264) show how complex the application of differentiation studies on a molecular basis will be. To interpret the cross-union experiments in terms of regulation of gene function will be indeed difficult, but from a neurochemical point of view rewarding.

IV. SUMMARY

This chapter intends to justify the growing interest of neurochemists
in problems of nerve-muscle cell relationships which was aroused especially by studies on changes of muscle metabolism induced in nerve cross-
union experiments. For this purpose not only an evaluation of experimental
and biochemical methods used in denervation and reinnervation experiments
but also a discussion of the main problems and data concerning nerve-muscle cell relationships seemed necessary. The evidence for the trophic
function of the nerve cell relatively independent of nerve-impulse activity
and neuromuscular transmission, the functional-metabolic correlations of
changes in nerve and muscle cell during denervation and reinnervation,
and their neurochemical implications are reviewed.

The optimal surgical methods and the possible experimental artifacts,
especially in nerve cross-union experiments, are discussed. The misin-
terpretations due to inadequate choice of animal, nerve, and type of muscle,
biochemical methods, and to side effects of denervation and reinnervation
studies are listed. Finally the contributions and limitations of denervation
and reinnervation experiments with respect to neurochemistry, especially
from a neurogenetic point of view, are discussed, the main intention being
to define conditions for "successful" reinnervation experiments. With
respect to the relatively new application of such studies to neurochemistry,
open questions in this field are considered.

REFERENCES

1. J. Zelená, "Development, Degeneration and Regeneration of Receptor
   Organs," in Progress in Brain Research, Vol. 13 (M. Singer and J. P.
   Schadé, eds.), Elseiver, Amsterdam, 1964, pp. 175-213.

2. A. F. W. Hughes, Aspects of Neural Ontogeny, Logos Press, Academic, New York, 1968.

3. G. Veneroni and M. R. Murray, Formation de novo and development
   of neuro-muscular junctions in vitro, J. Embryol. Exptl. Morphol.,
   21, 369 (1969).

4. D. W. James and R. L. Tresman, An electron-microscopic study of
   de novo formation of neuro-muscular junctions in tissue culture, Z.
   Zellforsch., 100, 126 (1969).

5. S. M. Crain, Bioelectric interactions between cultured fetal rodent
   spinal cord and skeletal muscle after innervation in vitro, J. Exptl.
   Zool., 173, 353 (1970).

6. M. Singer, The influence of the nerve in regeneration of the amphibian extremity, Quart. Rev. Biol., 27, 169 (1952).

7. E. Gutmann, Die funktionelle Regeneration der peripheren Nerven, Akademie-Verlag, Berlin, 1958.

8. E. Gutmann, "Neurotrophic Relations in the Regeneration Process," in Mechanisms of Neural Regeneration. Progress in Brain Research, Vol. 13 (M. Singer and J. P. Schadé, eds.), Elsevier, Amsterdam, 1964, pp. 72-112.

9. L. Guth, "Trophic" influences of nerve on muscle, Ann. Rev. Physiol., 48, 645 (1968).

10. R. Miledi, Properties of regenerating neuromuscular synapses in the frog, J. Physiol., 154, 190 (1960).

11. E. Gutmann and P. Hnik, "Denervation Studies in Research of Neurotrophic Relationships, in The Denervated Muscle (E. Gutmann, ed.), Czechosl. Acad. Sci., Prague, 1962, p. 13.

12. R. Miledi, "An Influence of Nerve Not Mediated by Impulses," in The Effect of Use and Disuse on Neuromuscular Functions (E. Gutmann and P. Hnik, eds.), Czechosl. Acad. Sci., Prague, 1963, p. 35.

13. L. Guth, "Trophic" effects of vertebrate neurons, Neurosci. Res. Progr. Bull., 7, 5 (1969).

14. B. Katz, The Release of Neural Transmitter Substances, Liverpool University Press, 1969.

15. D. B. Drachmann, Is acetylcholine the trophic neuromuscular transmitter? Arch. Neurol., 17, 206 (1967).

16. F. E. Bloom, L. L. Iversen, and F. O. Schmitt, Macromolecules in synaptic function, Neurosciences Res. Progr. Bull., 8, 329 (1970).

17. A. D. Smith, Summing up: some implications of the neuron as a secreting cell, Phil. Trans. Roy. Soc. London, B261, 423 (1971).

18. J. Musick and J. Hubbard, Release of proteins and acetylcholine at the neuromuscular junction, Proc. 24 Intern. Congr. I.U.P.S. Munich, 9, 410 (1971).

19. E. Scharrer, and B. Scharrer, "Neurosecretion," in Handbuch der mikroskopischen Anatomie des Menschen (W. von Möllendorf and W. Bargmann, eds.), Springer-Verlag, Berlin, 1964.

20. S. R. Cajal, Degeneration and Regeneration of the Nervous System, Oxford University Press, London, 1928.

21. J. Z. Young, Functional repair of nervous tissue, Physiol. Rev., 22, 318 (1942).

22. P. Weiss and H. B. Hiscoe, Experiments on the mechanism of nerve growth, J. Exptl. Zool., 107, 315 (1948).

23. L. Lubinska, "Axoplasmic Streaming in Regenerating and in Normal Nerve Fibers," in Progress in Brain Research, Vol. 13 (M. Singer and J. P. Schadé, eds.), Elsevier, Amsterdam, 1964, p. 1.

24. A. Dahlström, Axoplasmic transport (with particular respect to adrenergic neurons), Phil. Trans. Roy. Soc. London., B261, 325 (1971).

25. J. V. Luco and C. Eyzaguirre, Fibrillation and hypersensitivity to ACh in denervated muscle; effects of length of degenerating nerve fibers, J. Neurophysiol., 18, 65 (1955).

26. E. Gutmann, Z. Vodička, and J. Zelená, Changes in striated muscle after section of nerve, as a function of the length of the peripheral segment, Physiol. Bohemoslov., 4, 200 (1955).

27. C. R. Slater, Time course of failure of neuromuscular transmission after motor nerve section, Nature, 209, 305 (1966).

28. E. Gutmann, "Development and Maintenance of Neurotrophic Relations between Nerve and Muscle," in Ciba Foundation Symposium on Growth of the Nervous System (G. E. Wolstenholme, and M. O'Connor, eds.), Churchill, London, 1968, p. 233.

29. M. Singer, "A Theory of the Trophic Nervous Control of Amphibian Limb Regeneration, Including a Re-evaluation of Quantitative Nerve Requires," in Regeneration in Animals and Related Problems (V. Kiortsis and H. A. L. Trampusch, eds.), North-Holland, Amsterdam, 1965, pp. 20-32.

30. C. S. Thornton, Amphibian limb regeneration, Am. Zoologist, 10, 113 (1970).

31. A. J. Harris, S. Heinemann, D. Schubert, and H. Tarakis, Trophic interactions between cloned culture lines of nerve and muscle, Nature, 231, 298 (1971).

32. T. L. Lentz, Nerve trophic function: in vitro assay of effects of nerve tissue on muscle cholinesterase activity, Science, 171, 187 (1971).

33. M. C. Prestige, "Differentiation, Degeneration and the Role of the Periphery: Quantitative Considerations," in The Neurosciences (G. C. Quarton, T. Melnechuck, and G. Adelman, eds.), Rockefeller University Press, New York, 1970, p. 73.

34. J. Z. Young, Effects of use and disuse on nerve and muscle, Lancet, 1946-II, 109.

35. F. Nissl, Über die Veränderungen der Gaglienzellen am Facialis Kern Kaninchens nach Ausreissung des Nerven, Allg. Z. Psychiat., 48, 197 (1892).

36. E. G. Gray and R. W. Guillery, Synaptic morphology in the normal and degenerating nervous system, Intern Rev. Cytol., 19, 111 (1966).

37. S. O. Brattgard, J. E. Edström, and H. Hydén, The productive capacity of the neuron in retrograde reaction, Exptl. Cell Res., Suppl., 5, 185 (1958).

38. S. L. Palay and G. E. Palade, The fine structure of neurons, J. Biophys. Biochem. Cytol., 1, 69 (1965).

39. E. Gutmann, The trophic function of the nerve cell, Scientia, 63, 1 (1969).

40. S. Ochs, M. I. Sabri, and J. J. Johnson, Fast transport of labeled material in mammalian nerve, Science, 163, 686 (1969).

41. P. Lebowitz and M. Singer, Neurotrophic control of protein synthesis in the regenerating limb of the newt Triturus, Nature, 225, 824 (1970).

42. E. Gutmann and J. Z. Young, The reinnervation of muscle after various periods of atrophy, J. Anat., 78, 15 (1944).

43. E. J. Desmedt, The physio-pathology of neuromuscular transmission and the trophic influence of motor innervation, Am. J. Phys. Med., 38, 248 (1959).

44. R. Elul, R. Miledi, and E. Stefani, Neural control of contracture in slow muscle fibres of the frog, Acta Physiol. Latinoam., 20, 194 (1970).

45. J. N. Langley, Observations on denervated muscle, J. Physiol. (London), 50, 335 (1916).

46. A. G. Ginetzinski and N. M. Shamarina, Tonomotor phenomena in the denervated muscle (in Russian), Usp. Sovrem. Biol., 15, 283 (1942).

47. G. A. Kerkut, A. Shapira, and R. J. Walker, The transport of labeled material from CNS — along a nerve trunk, Comp. Biochem. Physiol., 23, 7 (1967).

48. I. M. Korr, P. N. Wilkinson, and F. W. Chornock, Axonal delivery of neuroplasmic components to muscle cells, Science, 153, 342 (1967).

49. A. J. Buller, J. C. Eccles, and R. M. Eccles, Interactions between motoneurons and muscles in respect of the characteristic speeds of their responses, J. Physiol. (London), 150, 417 (1960b).

50. A. J. Buller, W. F. H. M. Mommaerts, and K. Seraydarin, Enzymic properties of myosin in fast and slow twitch muscles of the cat following cross innervation, J. Physiol. (London), 205, 581 (1969).

51. Z. Drahota and E. Gutmann, Long-term regulatory influence of the nervous system on some metabolic differences in muscles of different function, Physiol. Bohemoslov., 12, 339 (1963).

52. G. Golisch, D. Pette, and H. Pichlmayer, Metabolic differentiation of rabbit skeletal muscle as induced by specific innervation, European J. Biochem., 16, 110 (1970).

53. G. Karpati and W. K. Engel, Transormation of the histochemical profile of skeletal muscle by "foreign" innervation, Nature, 215, 1509 (1967).

54. L. Guth, F. J. Samaha, and R. W. Albers, The neural regulation of some phenotypic differences between the fiber types of mammalian skeletal muscle, Exptl. Neurol., 26, 126 (1970).

55. I. Guth, "A Review of the Evidence for the Neural Regulation of Gene Expressions in Muscle," in Contractility of Muscle Cells and Related Processes (R. J. Podolsky, ed.), Prentice-Hall, Englewood Cliffs, N. J., 1971, p. 189.

56. M. Abercrombie and M. L. Johnson, The outwandering of cells in tissue cultures of nerves undergoing Wallerian degeneration, J. Exptl. Biol., 19, 266 (1942).

57. J. Takahashi, M. Nomura, and S. Furasawa, In vitro incorporation of C^{14}-amino acids into proteins of peripheral nerve during Wallerian degeneration, J. Neurochem., 7, 97 (1961).

58. S. Sunderland, Nerves and Nerve Injuries, E. S. Livingstone, Edinburgh and London, 1968.

59. E. Gutmann and R. Zák, Nervous regulation of nucleic acid level in cross-striated muscle. Changes in denervated muscle, Physiol. Bohemoslov., 10, 493 (1961).

60. E. D. Robert and J. T. Oester, Nerve impulses and trophic influences: Absence of fibrillation after prolonged conduction block, Arch. Neurol., 22, 20 (1970).

61. P. Weiss and H. Davis, Pressure block in nerves provided with arterial sleeves, J. Neurophysiol., 6, 269 (1943).

62. W. R. Merrington and P. W. Nathan, A study of post-ischaemic paraesthesia, J. Neurol. Neurosurg. Psychiat., 12, 1 (1949).

63. S. Thesleff, Supersensitivity of skeletal muscle produced by botulinum toxin, J. Physiol. (London), 151, 598 (1960a).

64. M. D. Sokoll, B. Sonesson, and S. Thesleff, Denervation changes produced in an innervated skeletal muscle by long-continued treatment with a local anesthetic, European J. Pharmacol., 4, 179 (1968).

65. P. Hnik, Motor function disturbances and excitability changes following deafferentation, Physiol. Bohemoslov., 5, 305 (1956).

66. A. Jefferson, Aspects of the segmental innervation of the cats hind limb, J. Comp. Neurol., 100, 569 (1954).

67. E. Gutmann and L. Guttmann, Factors affecting recovery of sensory function after nerve lesion, J. Neurol. Neurosurg. Psychiat., 5, 117 (1942).

68. I. Syrový, E. Gutmann, and J. Melichna, Differential response of myosin ATPase activity and contraction properties of fast and slow muscles following denervation, Experientia, 27, 1426 (1971).

69. L. Guth, Regeneration in the mammalian peripheral nervous system, Physiol. Rev., 36, 441 (1956).

70. E. Gutmann and F. K. Sanders, The recovery of fibre size and numbers during nerve regeneration, J. Physiol. (London), 101, 489 (1943).

71. F. K. Sanders and D. Whitteridge, Conduction velocity and myelin thickness in regenerating nerve fibres, J. Physiol., 105, 152 (1946).

72. J. Z. Young, "The Determination of the Specific Characteristics of Nerve Fibres," in Genetic neurology (P. Weiss, ed.), University of Chicago Press, 1950, p. 92.

73. E. Gutmann, Effect of delay of innervation on recovery of muscle after nerve lesions, J. Neurophysiol., 11, 279 (1948).

74. D. Duncan and W. H. Jarvis, Observation on repeated regeneration of the facial nerve in cats, J. Comp. Neurol., 79, 315 (1943).

75. J. Z. Young and P. B. Medawar, Fibrin suture of peripheral nerves, Lancet, 1940-II, 126.

76. E. Gutmann, L. Guttmann, P. B. Medawar, and J. Z. Young, The rate of regeneration of nerve, J. Exptl. Biol., 19, 14 (1942).

77. I. M. Tarlov, W. Boernstein, and D. Berman, Nerve regeneration. A comparative experimental study following suture by clot and thread, J. Neurosurg., 5, 62 (1948).

78. M. Singer, The combined use of fibrin film and clot in end-to-end union union of nerves. An experimental study, J. Neurosurg., 2, 102 (1945).

79. B. S. Freeman, Adhesive anastomosis technics for fine nerves: Experimental and clinical techniques, Am. J. Surg., 108, 529 (1964).

80. P. Weiss, Nerve regeneration in the rat following tubular spicing of several nerves, Arch. Surg. (Chicago), 46, 525 (1943).

81. P. Weiss, Nervous System (Neurogenesis). Analysis of Development (D. H. Willier, P. A. Weiss and V. Hamburger, eds.), Saunders, Philadelphia, 1955, p. 346.

82. R. W. Sperry, "Embryogenesis of Behavioral Nerve Nets," in Organogenesis (R. L. De Haan and H. Ursprung, eds.), New York, 1965, p. 161.

83. T. P. Feng, W. Y. Wu, and F. Y. Yang, Selective reinnervation of a "slow" or "fast" muscle by its original motor supply during regeneration of mixed nerves, Sceintia (Peking)., 14, 1717 (1965).

84. J. F. Y. Hoh, Selective reinnervation of fast-twitch and slow graded muscle fibres in the toad, Exptl. Neurol., 30, 263 (1971).

85. A. J. Buller, J. C. Eccles, and R. M. Eccles, Differentiation of fast and slow muscles in the cat hind limb, J. Physiol. (London), 150, 339 (1960a).

86. R. I. Close, Dynamic properties of fast and slow skeletal muscles of the rat during development, J. Physiol. (London), 173, 74 (1964).

87. E. Gutmann and I. Syrový, Metabolic differentiation of the anterior and posterior latissimus dorsi of the chicken during development, Physiol. Bohemoslov., 16, 232 (1967).

88. S. W. Kuffler and E. M. Vaughan Williams, Properties of the "slow" skeletal muscle fibres of the frog, J. Physiol. (London), 121, 318 (1953).

89. P. Krüger, Die Innervation phasisch bezw. tonisch reagierender Muskeln von Säugetieren und des Menschen, Acta Anat., 40, 186 (1960).

90. C. S. Sherrington, Some functional problems attaching to convergence, Proc. Roy. Soc. London, Series B, 105, 332 (1929).

91. W. K. Engel, The essentiality of histo- and cytochemical studies of skeletal muscle in the investigation of neuromuscular disease, Neurology, 12, 778 (1962).

92. V. Dubowitz and A. G. E. Pearse, A comparative histochemical study of oxidative enzyme and phosphorylase activity in skeletal muscle, Histochemie, 2, 105 (1960).

93. J. M. Stein, and H. A. Padykula, Histochemical classification of individual skeletal muscle fibres of the rat, Am. J. Anat., 110, 103 (1962).

94. G. F. Gauthier, "The Ultrastructure of Three Types in Mammalian Skeletal Muscle," in The Physiology and Biochemistry of Muscle as a Food (E. J. Briskey, R. G. Cassens, and B. B. Marsh, eds.), University of Wisconsin Press, Madison, 1970, p. 103.

95. S. Schiaffino, V. Hanzliková, and S. Pierbon, Relations between structure and function in rat skeletal muscle fibres, J. Cell Biol., 47, 107 (1970).

96. E. Henneman and C. B. Olson, Relations between structure and function in the design of skeletal muscles, J. Neurophysiol., 28, 581 (1965).

97. R. I. Close, Properties of motor units in fast and slow muscles of mammals, J. Physiol. (London), 193, 45 (1967).

98. R. E. Burke, D. N. Levine, F. E. Zajac, P. Tsairis, and W. K. Engel, Mammalian motor units, physiological-histochemical correlation in three types of cat gastrocnemius, Science, 174, 709 (1971).

99. J. C. Eccles, R. M. Eccles, and A. Lundberg, The action potentials of alpha motoneurones supplying fast and slow muscles, J. Physiol., 142, 275 (1958).

100. E. Henneman, G. Somjen, and D. O. Carpenter, Excitability and inhibitility of motoneurons of different sizes, J. Neurophysiol., 228, 599 (1965).

101. C. A. Elsberg, Experiments on motor nerve regeneration and the direct neurotization of paralyzed muscles by their own and by foreign nerves, Science, 45, 318 (1917).

102. S. Fex and I. Jirmanová, Innervation by nerve implants of "fast" and "slow" skeletal muscles of the rat, Acta Physiol. Scand., 76, 257 (1969).

103. E. Gutmann and V. Hanzliková, Effects of accessory nerve supply to muscle achieved by implantation into muscle during regeneration of its nerve, Physiol. Bohemoslov., 16, 244 (1967).

104. J. T. Aitken, Growth of nerve implants in voluntary muscke, J. Anat., 84, 38 (1950).

105. S. Fex, S. Sonesson, S. Thesleff, and J. Zelená, Nerve implants in botulinum poisoned mammalian muscle, J. Physiol. (London), 184, 872 (1966).

106. D. G. Gwan and J. T. Aitken, The formation of new motor endplates in mammalian skeletal muscle, J. Anat., 100, 111 (1966).

107. G. J. Romanes, Motor localization and the effects of nerve injury on the ventral horn cells of the spinal cord, J. Anat. (London), 80, 117 (1946).

108. I. A. Boyd and M. R. Dawey, Composition of Peripheral Nerves, E. S. Livingstone, London, 1968.

109. E. Gutmann and J. Zelená, "Morphological Changes in the Denervated Muscle," in The Denervated Muscle (E. Gutmann, ed.), Czechoslovak. Acad. Sci., Prague, 1962, p. 57.

110. P. Hnik, Muscle atrophy (in Czech), Babak-series, 43, 1966.

111. A. J. Harris and R. Miledi, A study of frog muscle maintained in organ culture, J. Physiol. (London), 221, 207 (1972).

112. R. Miledi and O. A. Throwell, Acetylcholine sensitivity of rat diaphragm maintained in organ culture, Nature, 194, 981 (1962).

113. D. M. Fambrough, Acetylcholine sensitivity of muscle fiber membranes. Mechanism of regulation by motorneurons, Science, 168, 372 (1970).

114. R. G. Harrison, Observations on the living developing nerve fibre, Proc. Soc. Exptl. Biol. (New York), 4, 140 (1907).

115. N. Robbins and T. Yonezawa, Developing neuromuscular junctions: First signs of chemical transmission during formationiin tissue culture, Science, 172, 395 (1971).

116. G. D. Fischbach, Synaptic potentials recorded in cell cultures of nerve and muscle, Science, 169, 1331 (1970).

117. M. Tolar and M. R. Murray, Nerve-muscle relations as studied in chick-neuromuscular model systems in vitro, Physiol. Bohemoslov. (in press).

118. E. Gutmann and J. Melichna, Contractile properties of different skeletal muscles of the rat during development, Physiol. Bohemoslov., 21, 1 (1972).

119. A. Bass, G. Lusch, and D. Pette, Postnatal differentiation of the enzyme activity pattern of energy-supplying metabolism in slow (red) and fast (white) muscles of chicken, European J. Biochem., 13, 289 (1970).

120. C. L. Vera and J. V. Luco, Synaptic transmission in spinal cord during Wallerian degeneration of dorsal root fibres, J. Neurophysiol., 21, 334 (1958).

121. P. G. Kostyuk, Functional presynaptic and postsynaptic changes during degeneration of central synapses, Federation Proc., 22, Suppl. T, 1101 (1963).

122. C. N. B. Dowman, J. C. Eccles, and A. K. McIntyre, Functional changes in chromatolysed motoneurons, J. Comp. Neurol., 98, 9 (1953).

123. J. E. Edström, Ribonucleic acid changes in the motorneurones of the frog during axon regeneration, J. Neurochem., 5, 43 (1959).

124. F. X. Albuquerque and R. J. Mc Issac, Fast and slow mammalian muscles after denervation, Exptl. Neurol., 26, 183 (1970).

125. P. Hnik and V. Skorpil, "Fibrillation Activity in Denervated Muscle," in The Denervated Muscle (E. Gutmann, ed.), Czechosovak. Acad. Sci, Prague, 1962, p. 135.

126. B. Salafsky, J. Bell, and M. A. Prewitt, Development of fibrillation potentials in denervated fast and slow skeletal muscles, Am. J. Physiol., 215, 637 (1968).

127. A. Sandow, Excitation-contraction coupling in skeletal muscle, Pharmacol. Rev., 17, 265 (1965).

128. E. Gutmann, Schnelle und langsame Muskeln, Ärztliche Forschung, 24, 157 (1970).

129. A. Hess, Vertebrate slow muscle fibres, Physiol. Rev., 56, 40 (1970).

130. C. H. Beatty and R. M. Boeck, "Biochemistry of the Red and White Muscle," in The Physiology and Biochemistry of Muscle as a Food (E. J. Briskey, R. G. Cassens, and B. B. Marsh, eds.), Univ. of Wisc. Press, Madison, 1970.

131. A. Bass, D. Brdiczka, P. Eyer, S. Hofer, and D. Pette, Metabolic differentiation of distinct muscle types at the level of enzymatic organization, European J. Biochem., 10, 198 (1969).

132. A. F. Huxley, Muscle, Ann. Rev. Physiol., 26, 131 (1964).

133. W. M. Paul, E. E. Daniel, C. M. Kay, G. Monckton, eds., Muscle, Pergamon Press, Symposium Publications Division, New York, 1964.

134. L. D. Peachy, Muscle, Ann. Rev. Physiol., 30, 401 (1968).

135. L. Ranvier, De quelques fait rélatifs à l'histologie et à la physiologie des muscles striés, Arch. Physiol. Norm et Path., 2, Sér. 1:5 (1874).

136. Chuen-Shang, Chung, Wu, Comparative studies on myosins from breast and leg muscle of chicken, Biochemistry, 8, 38 (1969).

137. J. Dow and A. Stracher, Identification of the essential light chains of myosin, Proc. Natl. Acad. Sci. U.S.A., 68, 1107 (1971).

138. M. H. Brooke, "Some Comments on Neural Influence on the Two Histochemical Types of Muscle Fibres," in The Physiology and Biochemistry of Muscle as a Food (E. J. Briskey, R. G. Cassens, and B. B. Marsh, eds.), Univ. of Wisc. Press, Madison, 1970, p. 131.

139. L. Guth, P. K. Watson, and W. C. Brown, Effects of cross-reinnervation on some chemical properties of red and white muscles of rat and cat, Exptl. Neurol., 20, 52 (1968).

140. N. Robbins, G. Karpati, and W. K. Engel, Hostochemical and contractile properties in the cross-innervated guinea pig soleus muscle, Arch. Neurol., 20, 318 (1969).

141. F. C. A. Romanul and J. P. Van der Meulen, Slow and fast muscles after cross innervations: Enzymatic and physiological changes, Arch. Neurol., 17, 387 (1967).

142. V. Edgerton, L. Gerchman, and R. Carrow, Histochemical changes in rat skeletal muscle after exercise, Exptl. Neurol., 24, 110 (1969).

143. G. Karpati and W. K. Engel, Neuronal trophic function, Arch. Neurol., 17, 542 (1967).

144. E. Gutmann, S. Schiaffino, and V. Hanzliková, Mechanism of compensatory hypertrophy in skeletal muscle of the rat, Exptl. Neurol., 31, 451 (1971).

145. L. Guth and H. Yellin, The dynamic nature of the so called "Fibre Types" of mammalian skeletal muscle, Exptl. Neurol., 31, 277 (1971).

146. E. C. B. Hall-Craggs, The contraction times and enzyme activity of two laryngeal muscles, J. Anat., 102, 241 (1968).

147. I. Syrový and E. Gutmann, ATPase activity of two rabbit laryngeal muscles, Experientia, 27, 248 (1971).

148. L. Edström and E. Kugelberg, Histochemical composition, distribution of fibres and fatiguability of single motor units, J. Neurol. Neurosurg. Psychiat., 31, 424 (1968).

149. E. Bajusz, "Red" skeletal muscle fibres: relative independence of neural control, Science, 145, 938 (1964).

150. G. Karpati and W. K. Engel, Histochemical investigation of fibre type ratios with the myofibrillar ATPase reaction in normal and denervated skeletal muscles of the guinea pig, Am. J. Anat., 122, 145 (1968).

151. M. A. Prewitt and B. Salafsky, Effect of cross innervation on biochemical characteristics of skeletal muscles, Am. J. Physiol., 213, 295 (1967).

152. D. M. Lewis, The effects of denervation on the speed of striated muscle, J. Physiol. (London), 161, 24P, (1962).

153. M. A. Prewitt and B. Salafsky, Enzymic and histochemical changes in fast and slow muscles after cross innervation, Am. J. Physiol., 218, 69 (1970).

154. E. L. Hogan, D. M. Dawson, and F. C. A. Romanul, Enzymatic changes in denervated muscle II. Biochemical studies, Arch. Neurol., 13, 274 (1965).

155. P. A. Weiss, "Neuronal Dynamics and Neuroplasmic Flow," in The Neurosciences (G. C. Quarton, T. Melnechuk, and G. Adelman, eds.), Rockefeller University Press, New York, 1970, p. 840.

156. B. G. Cragg, What is the signal for chromatolysis? Brain Res., 23, 1 (1970).

157. W. Hild, "Das Neuron," in Handbuch der mikroskopischen Anatomie des Menschen, Vol. IV/4, Springer-Verlag, Berlin, 1959, p. 1.

158. M. Cole, "Retrograde Degeneration of Axon and Soma in the Nervous System," in The Structure and Function of Nervous Tissue (G. H. Bourne, ed.), Vol. 1, 1968, p. 269.

159. E. Kugelberg and L. Edström, Differential histochemical effects of muscle contractions on phosphorylase and glycogen in various types of fibres: relation to fatigue, J. Neurol. Neurosurg. Psychiat., 31, 415 (1968).

160. R. E. Burke, Central nervous system control of fast and slow twitch motor units, 4th Intern. Congr. Electromyography, Brussels, 1971, p. 18.

161. J. F. Campa and W. K. Engel, Histochemistry of motor neurons and interneurons in the cat lumbar spinal cord, Neurology, 20, 559 (1971).

162. H. Hydén, "The neuron," in The Cell (J. Brachet-Mirski, ed.), Vol. IV, Academic, New York, 1960, p. 215.

163. J. Jarlsted, Functional location in the cerebellar cortex studied by quantitative determinations of Purkinyi cell RNA, Acta Physiol. Scand., 67, 243 (1966).

164. O. H. Lowry, The chemical study of single neurons, Harvey Lectures, series 58, Academic Press, New York, 1963, p. 7.

165. B. Droz, "Fate of Newly Synthesized Proteins in Neurons," in The Use of Radioautography in Investigating Protein Synthesis (C. P. Leblond and K. B. Warren, eds.), Academic, New York, 1965, p. 159.

166. B. Jakoubek and B. Semiginovský, The effect of increased functional activity on the protein metabolism of the nervous system, Intern. Rev. Neurobiol., 13, 255 (1970).

167. B. Jakoubek, E. Gutmann, J. Fischer, and A. Babický, Rate of protein renewal in spinal motorneurons of adolescent and old rats, J. Neurochem., 15, 633 (1968).

168. W. E. Watson, Observations on the nucleolar and total cell body nucleic acid of injured nerve cells, J. Physiol. (London), 196, 655 (1968).

169. M. J. Cohen, "Correlations between Structure, Function and RNA Metabolism in Central Neurons of Insects," in Invertebrate Nervous Systems (C. A. G. Wiersma, ed.), Univ. of Chicago Press, Chicago, 1967, p. 65.

170. B. Droz and C. P. Leblond, Axonal migration of proteins in the central nervous system and peripheral nerves as shown by radioautography, J. Comp. Neurol., 121, 325 (1963).

171. S. Ochs, "Fast Axoplasmic Transport of Proteins and Polypeptides in Mammalian Nerve Fibres, in Protein Metabolism of the Nervous System (A. Lajtha, ed.), Plenum, New York, 1970, p. 291.

172. B. Grafstein and M. Murray, Transport of protein in goldfish optic nerve during regeneration, Exptl. Neurol., 25, 494 (1969).

173. A. Edström and H. Mattson, Fast axonal transport in vitro in the sciatic system of the frog, J. Neurochem., 19, 205 (1972).

174. D. Bodian and R. C. Mellors, The regenerative cycle of motor neurons, with special reference to phosphatase activity, J. Exptl. Med., 81, 469 (1945).

175. W. E. Watson, Anaerobic glycolytic capacity of nerve cells after axotomy, J. Physiol. (London), 198, 77 (1968).

176. W. E. Watson, Some quantitative observations upon the oxidation of substrates of the tricarboxylic acid cycle in injured neurons, J. Neurochem., 13, 849 (1966).

177. I. Gersch and D. Bodian, Some chemical mechanisms in chromatolysis, J. Cellular Comp. Physiol., 21, 253 (1943).

178. E. Gutmann, B. Jakoubek, V. Rohlicek, and J. Skaloud, Effect of age on proteosynthesis in spinal motorneurones following nerve interruption as shown by histoautoradiography of ^{35}S methionine, Physiol. Bohemoslov., 11, 437 (1962).

179. R. P. Peterson, Cell size and rate of protein synthesis in ventral horn neurons, Science, 153, 1413 (1966).

180. B. Jakoubek, B. Semiginovský, and A. Dedicová, The effect of ACh on the synthesis of proteins in spinal motoneurons as studied by autoradiography, Brain Research, 25, 133 (1971).

181. E. Gutmann ed., The Denervated Muscle, Czechoslovak Acad. Sci., Prague, 1962.

182. U. Muscatello, A. Margreth, and M. Aloisi, On the differential response of sarcoplasm and myoplasm to denervation in from muscle, J. Cell Biol., 27, 1 (1965).

183. I. Hájek, E. Gutmann, M. Klicpera, and I. Syrový, The incorporation of S^{35} methionine into proteins of denervated and reinnervated muscle, Physiol. Bohemoslov., 15, 148 (1966).

184. R. Zák, J. Etlinger, and D. A. Fischman, "Studies on the Fractionation of Skeletal and Heart Muscle," in Modern Concepts in Muscle Development and the Spindle (R. Przybilski, J. P. Van Der Meulen, M. Victor, and B. Q. Banker, eds.), Elsevier, New York, 1971.

185. S. M. Heywood, R. M. Dowben, and A. Rich, A study of muscle polyribosomes and the co-precipitation of polyribosomes with myosin, Biochemistry, 7, 3289 (1968).

186. A. L. Goldberg, Protein turnover in skeletal muscle, J. Biol. Chem., 244, 3223 (1969).

187. K. L. Manchester and E. J. Harris, Effect of denervation on the synthesis of ribonucleic acid and deoxyribonucleic acid in rat diaphragm muscle, Biochem. J., 108, 188 (1968).

188. A. L. Goldberg and H. M. Goodman, Effects of disuse and denervation on amino acid transport by skeletal muscle, Am. J. Physiol., 216, 116 (1969).

189. R. C. Hider, E. B. Fern, and D. R. London, Relationship between intracellular amino acids and protein synthesis in the extensor digitorum longus muscle of rats, Biochem. J., 114, 171 (1969).

190. F. A. Sreter, J. C. Seidel, and J. Gergely, Studies on myosin from red and white skeletal muscles of the rabitt. I. Adenosine triphosphatase activity, J. Biol. Chem., 241, 5772 (1966).

191. M. Bárány, ATPase activity of myosin correlated with speed of muscle shortening, J. Gen. Physiol., 50, 197 (1967).

192. F. J. Samaha, L. Guth, and R. W. Albers, Differences between slow and fast muscle myosin: ATPase activity and release of associated proteins by p-chloromercuriphenylsulfonate, J. Biol. Chem., 245, 219 (1970).

193. S. V. Perry, The role of myosin in muscular contraction, Symp. Soc. Exptl. Biol., 22, 1 (1968).

194. D. A. Fischmann, "The Synthesis and Assembly of Myofibrils in Embryonic Muscle," in Current Topics in Developmental Biology, Academic, New York, 1970, p. 235-280.

195. F. H. M. Mommaerts, "The Role of the Innervation of the Functional Differentiation of Muscle," in The Physiology and Biochemistry of Muscle as a Food (E. J. Briskey, R. G. Cassens, and B. B. Marsh, eds.), Univ. of Wisc. Press, Madison, 1970, p. 53.

196. M. Bárány and R. I. Close, The transformation of myosin in cross-innervated rat muscles, J. Physiol. (London), 213, 455 (1971).

197. D. Pette, Plan and Muster im zellulären Stoffwechsel, Naturwissenschaften, 52, 597 (1965).

198. T. Bücher, W. Luh, and D. Pette, Hoppe-Seyler-Theirfelder Handbuch der physiologisch- und pathologischchemischen Analyse, Vol. 6/A, Springer-Verlag, Berlin, 1964, p. 292.

199. E. Gutmann, S. Tucek, and V. Hanzliková, Changes in cholinacetyltransferase and cholinesterase activities in the levator ani muscle of rats following castration, Physiol. Bohemoslov., 18, 195 (1969).

200. J. R. Florini and C. B. Breuer, Amino acid incorporation into proteins by cell free preparations from rat skeletal muscle, Biochemistry, 4, 253, 1965.

201. S. M. Heywood and A. Rich, In vitro synthesis of native myosin, actin and tropomyosin from embryonic chick polyribosomes, Proc. Natl. Acad. Sci. U.S.A., 59, 590 (1968).

202. H. Lövtrup-Rein, Synthesis of nuclear RNA in nerve and glial cells, J. Neurochem., 17, 853 (1970).

203. S. M. Heywood and M. Wagwu, Partial characterization of presumptive myosin messenger ribonucleic acid, Biochemistry, 8, 3839 (1969).

204. I. Hájek, M. Buresová, B. Jakoubek, Rapidly labeled RNA in rat diaphragm in vitro, Physiol. Bohemoslov., 21, 251 (1972).

205. U. E. Loening, The fractionation of high-molecular-weight ribonucleic acid by Polyacrylamide-Gel Electrophoresis, Biochem. J., 102, 251 (1967).

206. M. Jacob, J. Stevenin, R. Jund, C. Judes, and P. Mandel, Rapidly labelled ribonucleic acid in brain, J. Neurochem., 13, 619 (1966).

207. J. R. Tata, "Regulation of Protein Synthesis by Growth and Developmental Hormones," in Biochemical Actions of Hormones (G. Litwak, ed.), Academic, New York, 1970, p. 89.

208. J. R. Florini, Effects of testosterone on qualitative pattern of protein synthesis in skeletal muscle, Biochemistry, 9, 909 (1970).

209. M. Buresová and E. Gutmann, Effects of testosterone on protein synthesis and contractility of the levator ani muscle of the rat, J. Endocrinol., 50, 643 (1971).

210. R. Cihák, E. Gutmann, and V. Hanzliková, Involution and hormone-induced persistence of the M. sphincter (levator) ani in female rats, J. Anat., 106, 93 (1970).

211. M. Buresová, E. Gutmann, and V. Hanzliková, Differential effects of castration and denervation on protein synthesis in the levator ani muscle of the rat, J. Endocrinol., 54, 3 (1972).

212. O. Hudlická, "Nervous Control of Circulation in Skeletal Muscle," in Circulation in Skeletal Muscle (O. Hudlická, ed.), Pergamon Press, New York, 1968, p. 69.

213. E. Gutmann, M. Haniková, I. Hájek, M. Klicpera, and I. Syrový, The postdenervation hypertrophy of the diaphragm, Physiol. Bohemoslov., 15, 508 (1966).

214. B. Semiginovský, B. Jakoubek, M. Kraus, R. Erdöszová, Effect of the restraint stress on the oxygen consumption and ^{14}C leucine incorporation into the brain cortex slices in rats, Brain Res., 23, 298 (1970).

215. B. Jakoubek, B. Semiginovský, and A. Dedicová, The effect of ACTH on the synthesis of proteins in spinal motorneurones as studied by autoradiography, Brain Res., 25, 133 (1971).

216. C. D. Kochakian, "Regulation of Muscle Growth by Androgens," in The Physiology and Biochemistry of Muscle as a Food (E. J. Briskey, R. G. Cassens, and J. C. Trautman, eds.), Univ. of Wisc. Press, Madison, 1966, p. 81.

217. S. R. Hollán, Effect of nervous injury on the blood cell system. I, Acta Physiol. Hung., 12, 215 (1967).

218. S. Schiaffino and V. Hanzlikova, On the mechanism of compensatory hypertrophy in skeletal muscle, Experientia, 26, 152 (1970).

219. Ch. Chang and T. P. Feng, Comparative study on the proteins and nucleic acid changes of the anterior and posterior latissimus dorsi of the chicken following denervation, Acta Physiol. Sininica, 25, 312 (1962).

220. D. M. Stewart and A. W. Martin, Hypertrophy of the denervated hemidiaphragm, Am. J. Physiol., 186, 497 (1956).

221. O. M. Sola and A. W. Martin, Reaction of the diaphragm to denervation. Am. J. Physiol., 172, 324 (1953).

222. T. P. Feng and D. W. Lu, New lights of the phenomen of transient hypertrophy in the denervated hemidiaphragm of the rat, Scientia, 14, 1772 (1965).

223. I. Jirmanová and J. Zelená, Effect of denervation and tenotomy on slow and fast muscles of the chicken, Z. Zellforsch., 106, 333 (1970).

224. S. Fex, "Trophic" influences of implanted fast nerve on innervated slow muscle, Physiol. Bohemoslov., 18, 205 (1969).

225. E. Gutmann, The reinnervation of muscle by sensory nerve fibres, J. Anat. (London), 79, 1 (1945).

226. A. A. Zalewski, Effects of reinnervation on denervated skeletal muscle by axons of motor, sensory and sympathetic neurons, Am. J. Physiol., 219, 1675 (1970).

227. S. S. Tower, Trophic control of non-nervous tissues by the nervous system, J. Comp. Neurol., 67, 241 (1937).

228. J. C. Eccles, The Physiology of Synapses, Springer-Verlag, Berlin, 1964.

229. A. N. Studitskij, Experimental Surgery of Muscle (in Russian), Publ. House Acad. Sci. U.S.S.R., Moscow, 1956.

230. R. Miledi, The acetylcholine sensitivity of frog muscle fibers after complete or partial denervation, J. Physiol. (London), 151, 1 (1960).

231. S. Rose and P. H. Glow, Denervation effects on the presumed de novo synthesis of muscle cholinesterase and the effects of acetylcholine availability on retinal cholinesterase, Exptl. Neurol., 18, 267 (1967).

232. B. C. Goodwin and I. W. Sizer, Effects of spinal cord and substrate on acetylcholinesterase in chick embryonic skeletal muscle, Develop. Biol., 11, 136 (1965).

233. R. Birks, B. Katz, and R. Miledi, Physiological and structural changes at the amphibian myoneural junction, in the course of nerve degeneration, J. Physiol. (London), 150, 145 (1960).

234. G. Vrbová, The effect of motorneurone activity on the speed of contraction of striated muscle, J. Physiol. (London) (1963).

235. S. Salmons and G. Vrbová, The influence of activity on some contractile characteristics of mammalian fast and slow muscles, J. Physiol. (London), 201, 535 (1969).

236. C. A. Villee, Differentiation and enzymatic heterogeneity, Federation Proc., 25, 874 (1966).

237. I. Jirmanová, P. Hnik, and J. Zelená, Implantation of "fast" nerve into slow muscle in young chicken, Physiol. Bohemoslov., 20, 199 (1971).

238. R. I. Close, Dynamic properties of fast and slow skeletal muscles of the rat after nerve cross union, J. Physiol. (London), 204, 331 (1969).

239. F. J. Samaha, L. Guth, and R. W. Albers, The neural regulation of gene expression in the muscle cell, Exptl. Neurol., 27, 276 (1970).

240. P. Weiss, "An Introduction to Genetic Neurology, " in Genetic Neurology (P. Weiss, ed.), The Univ. Chicago Press, Chicago, 1950, p. 1.

241. R. L. Sidman, "Cell Proliferation, Migration and Interaction in the Developing Mammalian Control Nervous System, " in The Neurosciences (F. O. Schmitt, ed.), Rockefeller Univ. Press, New York, 1970, p. 160.

242. M. Jacobson, "Development, Specification and Diversification of Neuronal Connections, " in The Neurosciences (F. O. Schmitt, ed.), Rockefeller Univ. Press, New York, 1970, p. 116.

243. R. L. Davidson, Regulation of differentiation in cell hybrids, Federation Proc., 30, 926 (1971).

244. P. Hnik, I. Jirmanová, L. Vyklický, J. Zelená, Fast and slow muscles of the chick after nerve cross-union, J. Physiol. (London), 193, 325 (1967).

245. J. B. Peter, R. N. Jeffres, and D. R. Lamb, Exercise: effects on hexokinase activity in red and white skeletal muscle, Science, 160, (1968).

246. J. Nolte and D. Pette, "Quantitative Microscope-Photometric Determination of Enzyme Activities in Cryostat Sections," in Recent Advances in Quantitative Histo- and Cytochemistry, Hans Huber Publishers, Bern, Stuttgart, Vienna, 1971, p. 54.

247. R. H. Locker and C. J. Hagyard, Variation of the small subunits in different myosins, Arch. Biochem. Biophys., 122 (1967).

248. W. M. Kuehl and R. S. Adelstein, The absence of 3-methylhistidine in red, cardiac and fetal myosins, Biochem. Biophys. Res. Commun., 39, 956 (1970).

249. M. Bárány and K. Bárány, Adenosine triphosphate-dependent reaction of 1-fluoro-2,4-dinitrobenzene with various myosins, J. Biol. Chem., 244, 5206 (1969).

250. J. E. Ebert, "Molecular and Cellular Interactions in Development," in The Neurosciences (H. Quarton and F. O. Schmitt, eds.), The Rockefeller University Press, New York, 1967, p. 241.

251. G. M. Tomkins, T. D. Gelehrter, D. Granner, D. Martin, H. H. Samuels, and E. B. Thompson, Control of specific gene expression in higher organs, Science, 166, 1474 (1969).

252. F. Jacob and J. Monod, Genetic regulatory mechanisms in the synthesis of proteins, J. Mol. Biol., 3, 318 (1961).

253. R. P. Zhenevskaya, Experimental histological investigation of striated muscle tissue, Rev. Can. Biol., 21, 457 (1962).

254. N. V. Kchromov-Borisov and M. J. Michelson, The mutual disposition of cholinoreceptors of locomotor muscles and the changes in their disposition in the course of evolution, Pharmacol. Rev., 18, (3) 1051 (1966).

255. W. Grampp, J. B. Harris, and S. Thesleff, Inhibition of denervation changes in mammalian skeletal muscle, J. Physiol. (London), 217, 47 (1971).

256. I. Syrový, I. Hájek, and E. Gutmann, Degradation of proteins of M. latissiumus dorsi anterior and posterior of the chicken, Physiol. Bohemoslov., 14, 17 (1965).

257. P. Biron, J. C. Dreyfus, and F. Schapira, Différences métabolic entre les muscles rouges et blancs chez le lapin, Compt. Rend. Soc. Biol. (Paris), 158, 1841 (1964).

258. E. Gutmann, V. Hanzliková, and E. Holecková, Development of fast and slow muscles of the chicken in vivo and their latent period in tissue culture, Exptl. Cell. Res., 56, 33 (1969).

259. R. Zák, D. Grove, and M. Rabinowitz, DNA synthesis in the rat diaphragm as an early response to denervation, Am. J. Physiol., 216, 647 (1969).

260. J. R. Florini and F. L. Dankberg, Changes in ribonucleic acid and protein synthesis during induced cardiac hypertrophy, Biochemistry, 10, 530 (1971).

261. E. Gutmann and P. Hnik, Effect of distance from nerve cell on protein synthesis in spinal neurons after nerve section (unpublished, 1968).

262. E. Gutmann, Open questions in the study of the "trophic" function of the nerve cell, Topical Probl. Phychiat. Neurol., 10, 54 (1970).

263. J. P. Changeux and T. R. Podleski, "Remarks on the cholinergic Receptors of the Electroplax Membrane," in Macromolecules. Biosynthesis and Function, FEBS Symposium, Vol. 21, Academic, New York, 1970, p. 329.

264. H. Ursprung, W. J. Dickinson, G. Murison, and W. H. Sofer, Developmental Enzymology, FEBS Symposium, 21, Academic Press, New York, 1970, p. 231.

NOTE ADDED IN PROOF

Since this work went to press, some papers have been published pertinent to the problem of nerve-muscle relationship and indicative of trends and controversial aspects of present research.

(a) The question whether neural control of muscle is primarily mediated by nerve impulse activity or by neurotrophic agents transported by axoplasmic flow has been brought back into the foreground by "dissociation" experiments, in which either a prolonged blockade of nerve impulses only were produced by local anaesthetic implants (265) or a suppression of axoplasmic flow related to damage of microtubules was induced by colchicin without affecting impulse activity (266). In both cases, endplates (59, 265, 266) and spontaneous transmitter release remained intact, but the entire muscle membrane became sensitive to ACh. In the first type of experiment, the spread of ACh sensitivity, thought to be a sign of denervation, could be prevented by stimulation of the nerve below the block, and it was concluded that impulse activity controls chemosensitivity of the muscle membrane (265, 267, 268). The second type of experiment affected intra-axonal flow of neurotrophic agents without affecting the membrane function or conduction of nerve impulses and strengthens the evidence for neurotrophic control related to axoplasmic flow. The "dissociation" experiments emphasize the differentiation of the two main types of nervous control, i.e., by impulse activity and by neurotrophic agents dependent on intra-axonal flow. However, they also show that the appearance of extrajunctional ACh sensitivity is not an unequivocal sign of denervation and test of neurotrophic control respectively. During myogenesis some myoblasts and all myotubes are sensitive to ACh (269, 270) but chemosensitivity is apparently secondarily influenced by innervation (10, 271). The mechanism of this control, however, is still not clear.

(b) Factors affecting fast transport of material in nerve fibers (272) are studied, and further suggestions of transfer of "trophic" substances in the central (273) and peripheral (274) nervous system indicate the importance of axoplasmic flow for neurotrophic control.

(c) Nerve cross-union studies have shown that ultrastructure of slow, multiply innervated muscle fibers becomes altered toward the fast type if nerve cross-union is performed at an early stage of development (275). However, an ultrastructural transformation of slow to fast muscle fibers has been observed also after tenotomy (276). A multiple control of muscle has to be assumed; this is also shown by a study of differentiation of neural and hormonal influences affecting muscle fiber pattern after cross-union of nerves to a hormone-sensitive muscle (277).

(d) Studies on differentiation of muscles and muscle fibers have been intensively reviewed (278, 279). A wider range of different parameters of single muscle fibers from fast and slow muscles has been demonstrated; however, the main differentiation in contraction time, membrane properties, and ultrastructural features was confirmed (280). A change of pattern of activity was shown to change activities of key enzymes of energy-supplying metabolism and related muscle fiber pattern (281). Further studies on metabolic profiles of three main types of skeletal muscle recognize "fast-twitch-red," "fast-twitch-white," and "slow-twitch intermediate" fibers (282). The question of differentiation of three muscle fiber types is studied intensively (278). However, it appears that the basic distinction of fast and slow muscle fibers, the fast ones differing according to preferential glycolytic and oxidative energy supply metabolism respectively, is still the most useful one.

(e) Transplantation with related degeneration and regeneration of muscle (229, 283) is increasingly utilized in studies of nerve muscle relationship. Grafts of minced muscle tissue and free grafts are used (284, 285).

Almost full functional recovery of regenerated muscle is observed (286). The muscle changes from a slow to a fast type as during ontogenetic development (287). Functional and histochemical conversion of heterotopically grafted muscle occurs according to the type of innervation (288, 289). Fast regeneration of muscle is observed after use of myotoxic anesthetics which leave the nerve intact (270). Transplantation of muscle promises to be a potent tool in studies of neuronal specificity, and differentiation of myogenic and neurogenic mechanisms in myogenesis and maintenance of muscle.

REFERENCES (Continued)

265. T. Lomo and J. Rosenthal, Control of ACh sensitivity by muscle in the rat, J. Physiol., 221 (1972).

266. E. X. Albuquerque, J. E. Warnick, J. R. Tasse, and F. M. Sansone, Effects of vinablastine and colchicine on neural regulation of the fast and slow skeletal muscles of the rat: An electrophysiological and ultrastructural study, J. Gen. Physiol. (in press).

267. R. Jones and G. Vrbová, Can denervation hypersensitivity be prevented, J. Physiol., 217, 67pp., (1971).

268. D. B. Drachman and F. Witzke, Trophic regulation of acetylcholine sensitivity of muscle: Effect of electrical stimulation, Science, 176, 514 (1972).

269. D. Fambrough and J. E. Rash, Development of acetylcholine sensitivity during myogenesis, Develop. Biol., 26, 55 (1971).

270. I. Jirmanová and S. Thesleff, Ultrastructural study of experimental muscle degeneration and regeneration in the adult rat, Z. Zellforsch., 131, 77 (1972).

271. S. Thesleff, Effects of motor innervation on the chemical sensitivity of skeletal muscle, Physiol. Rev., 40, 734 (1960).

272. S. Ochs, Fast transport of material in mammalian nerve fibres, Science, 176, 252 (1972).

273. B. Grafstein, Transneuronal transfer of radioactivity in the central nervous system, Science, 172, 177 (1971).

274. J. Alvarez and M. Püschel, Transfer of material from efferent axons to sensory epithelium in the goldfish vestibular system, Brain Res., 37, 265 (1972).

275. J. Zelená and I. Jirmanová, Ultrastructure of chicken slow muscle after nerve cross union, Exptl. Neurol. (in press).

276. R. S. Hikida, Morphological transformation of slow to fast muscle fibres after tenotomy, Exptl. Neurol., 35, 265 (1972).

277. V. Hanzlíková and E. Gutmann, Effect of foreign innervation on the androgen-sensitive levator ani muscle of the rat, Z. Zellforsch., 135, 165 (1972).

278. R. I. Close, Dynamic properties of mammalian skeletal muscles, Physiol. Rev., 52, 129 (1972).

279. J. Zachar, Electrogenesis and Contractility in Skeletal Muscle Cells, University Park Press, Baltimore, 1971.

280. A. R. Luff and H. L. Atwood, Membrane properties and contraction of single muscle fibres in the mouse, Am. J. Physiol., 222, 1435 (1972).

281. D. Pette, H. W. Staudte, and G. Vrbová, Physiological and biochemical changes induced by long-term stimulation of fast muscle, Die Naturwissenschaften, 59, 469 (1972).

282. J. B. Peter, R. J. Barnard, V. R. Edgerton, C. A. Gillespeie, and K. E. Stempel, Metabolic profiles of three fibre types of skeletal muscle in guinea pig and rabbits, Biochemistry, 11, 2627 (1972).

283. B. M. Carlson, The regeneration of minced muscles, Monographs in Developmental Biology, Vol. 4, S. Karger, Basel, (1972).

284. N. Thompson, Investigation of autogeneous skeletal muscle free grafts in the dog, Transplantation, 12, 353 (1971).

285. B. Carlson, The functional morphology of regenerating and transplanted mammalian muscles, J. Physiol. (in press).

286. B. Salafsky, Functional studies of regenerated muscles from normal and dystrophic mice, Nature, 229, 94 (1971).

287. B. M. Carlson and E. Gutmann, Development of contractile properties of minced muscle regenerates in the rat, Exptl. Neurol., 36, 239 (1972).

288. E. Gutmann, V. Hanzlíková, and J. Štichová, Functional and histochemical properties of the transplanted levator ani muscle of the rat, Physiol. Bohemoslov. (in press).

289. B. Salafsky, Studies on the physiology of skeletal muscle regenerates, J. Physiol. (in press).

290. A. J. Buller and D. M. Lewis, Further observations on mammalian cross-innervated muscle, J. Physiol., 178, 343 (1965).

AUTHOR INDEX

Numbers in parentheses are reference numbers and indicate that an author's work is referred to although his name is not cited in the text. Underlined numbers give the page on which the complete reference is listed.

A

Abbas, T.M., 3(13), 43
Abdel-Latif, A.A., 148(12, 14, 20, 21), 150(29), 155(29, 33, 35a), 158(12, 14), 161-164(14, 20, 29, 52, 67, 71, 72, 74, 80), 166(20, 21, 74), 167(20, 74, 81), 171(67), 172(21), 173(21), 174(20, 74, 98a), 175(98a), 176(98a), 181-187
Abercrombie, M., 194(56), 236
Abood, L.G., 162(60), 163(63, 65, 67, 71), 171(65, 67), 185, 186
Abt, K.R., 4(26), 44
Adams, D.W., 4(41), 45
Adelstein, R.S., 208(248), 228(248), 250
Agrawal, H.C., 99(85), 110
Aguero, O., 41(199), 56
Aguilar, J., 163(66), 171(66), 185
Ahmad, J., 158(42), 183
Aitken, J.T., 202(104, 106), 239
Albers, R.W., 148(10), 158(10), 159(10), 162(10), 181, 193(54), 206(54), 208(54, 192), 218(192), 225-227(54, 239), 236, 246, 249
Albuquerque, F.X., 205(124), 241
Aleem, F.A., 4(27), 44
Alivisatos, S.G.A., 155(35a), 183
Aloisi, M., 217(182), 245
Alvarez, H., 4(37), 44

Amarose, A.P., 5(60), 46
Ametani, T., 119(19), 141
Anderson, N.H., 89(53), 108
Anderson, R.L., 20(123), 51
Anderson, W.R., 4(44), 45
Andrease, U., 21(138), 52
Andreass, J.L., 99(109), 112
Angermeir, W.F., 99(86), 110
Annau, Z., 90(61), 108
Appel, J.B., 91(65), 99(101, 115, 131), 109, 111, 112, 114
Appel, S.H., 148(22), 150(30), 163(76), 164(22), 166(22), 171(68), 173, 181, 182, 185, 186
Appleman, F., 8(68), 46
Armstrong, D., 18(113, 115), 21(126), 50, 51
Arnaiz, G., 155(34), 163(34), 183
Arnaiz, R., 148(10), 158(10), 159(10), 162(10), 181
Arnott, H.J., 118(12), 141
Aronson, S.M., 21(132), 51
Askari, A., 174(97), 187
Atkins, L., 3(4), 4(4), 42
Atkinson, A., 161, 162(54), 184
Austin, J., 18(111, 113, 115), 21(125, 126), 50, 51
Austin, L., 163(75), 186
Autilio, L.A., 163(76), 186
Avigan, J., 26(155), 53
Axelrod, J., 120(37), 142
A'Zary, E., 11(83), 47
Azrin, N.H., 99(104, 122), 112, 113

Other books of interest to you...

Because of your interest in our books, we have included the following catalog of books for your convenience.

Any of these books are available on an approval basis. This section has been reprinted in full from our *medicine/drugs* catalog.

If you wish to receive a complete catalog of MDI books, journals and encyclopedias, please write to us and we will be happy to send you one.

MARCEL DEKKER, INC.
95 Madison Avenue, New York, N.Y. 10016

medicine
including immunology, pathology, physiology, and virology

drugs
including medicinal chemistry, pharmacology, drug abuse, and pharmaceutical chemistry

BIEL and ABOOD *Biogenic Amines and Physiological Membranes in Drug Therapy*

In 2 Parts

(Medicinal Research Series, Volume 5) edited by JOHN H. BIEL, *Abbott Laboratories, North Chicago, Illinois,* and LEO G. ABOOD, *University of Rochester, New York*

Part A 176 pages, illustrated. 1971
Part B 384 pages, illustrated. 1971

Details pertinent developments that contribute to the understanding of the excitatory membrane and the mechanisms of action of biogenic amines. Relevance to the clinical situation and to the etiology of the disease process is stressed. Of importance to the research scientist working in physiology, pharmacology, and biochemical pharmacology, as well as to the medical student and physician in clinical pharmacology, especially in the areas of neurology, cardiovascular disease, and mental health.

CONTENTS:

Part A: The role of proteins and lipids in membrane structure and function, *L. G. Abood and A. Matsubara.* Biophysical role of Na^+, K^+, Mg^{2+}-activated adenosine triphosphatase in nerve cell membrane, *R. Tanaka.* Nerve excitation and "phase transition" in membrane macromolecules, *I. Tasaki.* Divalent cation, organic cation, and polycation interaction with excitable, thin lipid membranes, *R. C. Bean, W. C. Shepherd, and J. T. Eichner.*

Part B: Histamine and its role in biological systems, *R. W. Schayer.* Dopamine: Its physiology, pharmacology, and pathological neurochemistry, *O. Hornykiewicz.* The uptake of biogenic amines, *L. L. Iversen.* False neurotransmitters in the mechanism of action of central nervous system and cardiovascular drugs, *I. J. Kopin.* The application of biochemical techniques in the search for drugs affecting biogenic amines, *C. R. Creveling and J. W. Daly.* The "NIH shift" and a mechanism of enzymatic oxygenation, *D. M. Jerina, J. W. Daly, and B. Witkop.*

BURGER *Drugs Affecting the Central Nervous System*

(Medicinal Research Series, Volume 2) edited by ALFRED BURGER, *Department of Chemistry, University of Virginia, Charlottesville*

456 pages, illustrated. 1968

CONTENTS: Mechanisms of narcosis, *E. R. Larsen, R. A. Van Dyke, and M. B. Chenoweth.* The chemical anatomy of potent morphine-like analgesics, *P. A. J. Janssen and C. A. M. Van der Eycken.* Chemopharmacologic approaches to the treatment of mental depression, *J. H. Biel.* The psychotomimetic glycolate esters, *L. G. Abood.* Psychotomimetic agents, *A. Hofmann.* Structure-activity relationships in the 1,4-benzodiazepine series, *L. H. Sternbach, L. O. Randall, R. Banziger, and H. Lehr.* Synthetic centrally acting skeletal muscle relaxants, *H. B. Donahoe and K. K. Kimura.* Substituted phenothiazines: pharmacology and chemical structure, *E. F. Domino, R. D. Hudson, and G. Zografi.*

BURGER *Drugs Affecting the Peripheral Nervous System*

(Medicinal Research Series, Volume 1) edited by ALFRED BURGER, *Department of Chemistry, University of Virginia, Charlottesville.*

648 pages, illustrated. 1967

CONTENTS: Molecular aspects of cholinergic mechanisms, *S. Ehrenpreis.* Postganglionic parasympathetic stimulants (muscarinic drugs), *H. L. Friedman.* Postganglionic parasympathetic depressants (cholinolytic or atropinelike agents), *J. G. Cannon and J. P. Long.* Ganglionic stimulant and depressant agents, *L. Gyermek.* Drugs acting at nerve-skeletal-muscle junctions, *J. J. Lewis and T. C. Muir.* Reversible inhibitors of cholinesterase, *J. P. Long and C. J. Evans.* Acid-transferring inhibitors of acetylcholinesterase, *I. B. Wilson.* Sympathomimetic (adrenergic) stimulants, *A. M. Lands and T. G. Brown, Jr.* Synthetic postganglionic sympathetic depressants, *N. B. Chapman and J. D. P. Graham.* Effects of drugs on the afferent nervous systems, *C. M. Smith.*

BURGER Selected Pharmacological Testing Methods

(Medicinal Research Series, Volume 3)

edited by ALFRED BURGER, *Department of Chemistry, University of Virginia, Charlottesville.*

536 pages, illustrated. 1968

CONTENTS: The philosophy of pharmacological testing, *K. K. Chen.* Biostatistics in pharmacological testing, *C. W. Dunnett.* Significance of stimulation and inhibition of drug metabolism in pharmacological testing, *G. J. Mannering.* Methods for appraisal of analgetic drugs for addiction liability, *J. Cochin.* Testing for antihypertensive drugs, *W. Freyburger.* Evaluation of drugs affecting the contractility and electrical properties of the heart, *B. Katzung.* Ganglionic blocking agents, *J. A. Bevan.* Evaluation of liver function methodology, *G. L. Plaa.* Evaluation of diuretic agents, *B. R. Nechay.* Evaluation of drugs against effects of radiation, *P. S. Timiras and A. Vernadakis.* Testing of drugs for therapeutic potential in Parkinson's disease, *D. J. Jenden.* Evaluation of antitussive agents, *Y. Kasé.* Evaluation of sedative—hypnotics in the course of psychopharmacological testing, *R. F. Tislow.*

CHATTEN Pharmaceutical Chemistry

In 2 Volumes

edited by LESLIE G. CHATTEN, *Faculty of Pharmacy, University of Alberta, Edmonton*

Vol. 1 Theory and Application
520 pages, illustrated. 1966

Vol. 2 Instrumental Techniques
792 pages, illustrated. 1969

This new and comprehensive two volume textbook in pharmaceutical chemistry is for use of the senior undergraduate and graduate student studying analytical chemistry and instrumental techniques as applied to chemistry

"It will be welcomed by workers engaged in pharmaceutical analysis and should find a place in libraries of all colleges of pharmacy and medicine."—F. Kurzer, *Chemistry in Britain.* April, 1967

Volume 1: Introduction and technique, *M. Pernarowski.* Gravimetric analysis, *R. T. Coutts.* Acid-base titrations and pH, *J. W. Steele.* Precipitation, complex formation, and oxidation—reduction methods, *J. A. Zapotocky.* Acidimetry and alkalimetry, *M. I. Blake.* Nonaqueous titrimetry, *L. G. Chatten.* Complexometric titrations, *J. P. Leyda.* Alkaloidal assay and crude drug analysis, *W. C. Evans.* Miscellaneous methods, *J. E. Sinsheimer.* Ion exchange, *M. C. Vincent.* Column, thin-layer, and paper chromatography, *J. C. Morrison.* Analysis of fixed oils, fats, and waxes, *A. C. Glasser.* Analyses of volatile oils, *I. C. Nigam.*

Volume 2: Absorption spectrophotometry, *M. Pernarowski.* Infrared spectroscopy, *R. Coutts.* Raman spectroscopy, *E. A. Robinson and D. S. Lavery.* Fluorometry, *D. E. Guttman.* Atomic absorption spectroscopy, *J. W. Robinson.* Mass spectrometry, *A. Chisholm.* Nuclear magnetic resonance spectroscopy, *M. P. Mertes.* Turbidimetry: Nephelometry; colloidimetry, *F. T. Semeniuk.* Optical crystallography, *J. A. Biles.* X-ray analysis, *J. W. Shell.* Refractometry, *R. A. Locock.* Polarimetry, *R. A. Locock.* Potentiometric titrations, *J. G. Jeffrey.* Current flow methods, *S. Eriksen.* Coulometric methods and chronopotentiometry, *P. Kabasakalian.* Polarography, *F. W. Teare.* Amperometric titrations, *F. W. Teare.* Gas chromatography, *C. A. Bliss.* Radiochemical techniques, *A. Noujaim.*

DI CARLO Drug Metabolism Reviews

(Book Edition)

edited by FREDERICK J. DICARLO, *Department of Drug Metabolism, Warner-Lambert Research Institute, Morris Plains, New Jersey*

Vol. 1 366 pages, illustrated. 1973

Presents critical, in-depth reviews on the many aspects of drug metabolism, including discussions on the absorption, biotransformation, and excretion of drugs. Of fundamental concern to medicinal chemists, toxicologists, pharmacologists, microbiologists, pharmacokineticists, immunologists, and any other scientist involved in drug design. Also valuable reading for analytical chemists, mass spectroscopists, and enzymologists.

CONTENTS: Quantitative relationships between lipophilic character and drug metabolism, *C. Hansch.* Biopharmaceutical considerations in drug formulation design and evaluation, *S. Kaplan.* The pharmacodynamics of drug interaction, *D. Yesair, F. Bullock, and J. Coffey.* Metabolism of hydrazine derivatives of pharmacologic interest, *M. Juchau and A. Horita.* Recent developments in beta adrenergic blocking drugs, *K. Wong and E. Schreiber.* Metabolism of carbamate insecticides, *T. Fukuto.* Microsomal drug metabolizing enzymes in insects, *C. Wilkinson and L. Brattsten.* Effects of environmental organophosphorous insecticides on drug metabolism, *R. Stitzel, J. Stevens, and J. McPhillips.* Metabolic fate of hexobarbital, *M. Bush and W. Weller.* Benzodiazepine metabolism in vitro, *S. Garattini, F. Marcucci, and E. Mussini.*

EINSTEIN Methadone Maintenance

edited by STANLEY EINSTEIN, *Division of Drug Addiction, New Jersey College of Medicine and Dentistry*

264 pages, illustrated. 1971

Describes a variety of experiences in the United States and Canada in treating narcotic addicts with a new therapeutic technique, and represents the first compilation focusing exclusively on methadone treatment. An essential book for those working in the field of methadone maintenance; community leaders seeking guidelines for the development of such programs; and health professionals and laymen involved in the problem of drug abuse.

CONTENTS: Treatment of narcotic addicts in New York City, *R. E. Trussell.* Research on methadone maintenance treatment, *V. P. Dole.* Further experience with methadone in the treatment of narcotics users, *J. H. Jaffe.* Methadone maintenance programs in Minneapolis, *R. Maslansky.* Methadone maintenance in St. Louis, *R. Knowles, S. Lahiri, and G. Anderson.* The man alive program, *E. P. Davis.* Two methods of utilizing methadone in the outpatient treatment of narcotic addicts, *W. F. Wieland and C. D. Chambers.* Low and high methadone maintenance in the out-patient treatment of the hard core heroin addict, *H. R. Williams.* The New Haven methadone maintenance program, *H. D. Kleber.* Methadone in New Orleans: Patients, problems, and police, *W. A. Bloom, Jr. and E. W. Sudderth.* Methadone related deaths in New York City, *M. M. Baden.* Methadone in Miami, *J. Hoogerbeets.* Blockade with methadone, cyclazocine, and naloxone, *A. M. Freedman, A. Zaks, R. Resnick, and M. Fink.* Evaluation of methadone maintenance treatment program, *F. R. Gearing.* Commentary on the second national conference on methadone treatment, *W. R. Martin.* A bibliography of the methadone treatment of heroin addiction, *J. Langrod.*

FARBER   The Pathology of Transcription and Translation

(The Biochemistry of Disease Series, Volume 2)

edited by EMMANUEL FARBER, *Fels Research Institute, Temple University School of Medicine, Philadelphia, Pennsylvania*

192 pages, illustrated. 1972

Presents the first systematic description of how interference with the synthesis and metabolism of DNA, RNA, and protein is related to disease in higher organisms, including man. Designed to bridge the gap between the advances being made in molecular and cellular biology and human disease. Of vital interest to pathologists, biochemists, biologists, clinical chemists, medical students, graduate students in biology, physicians, and others involved in medical research.

CONTENTS: Pathology of DNA, *R. Baserga.* The control of gene expression, *G. Stein.* DNA metabolism, cell death, and cancer chemotherapy, *M. W. Lieberman.* Pathology of RNA, *J. Farber.* Response of nucleus and nucleolus to inhibition of RNA synthesis, *H. Shinozuka.* Pathogenesis of the cellular lesions produced by α-amanitin, *L. Fiume.* The pathology of translation, *E. Farber.*

FEATHERSTONE   A Guide to Molecular Pharmacology-Toxicology

In 2 Parts

(Molecular Pharmacology Series)

edited by R. M. FEATHERSTONE, *Department of Pharmacology, University of California, School of Medicine, San Francisco*

in preparation. 1973

Examines basic mechanisms underlying pharmacology-toxicology and presents the newer methods and approaches to the field. For postgraduate students in pharmacology, biochemistry, physiology, pharmaceutical chemistry, toxicology, and environmental science.

CONTENTS:

Part I: The use of models of the cell membrane in determining the mechanism of drug action, *L. Mullins.* Membrane transport receptors for monosaccharides, *M. Silverman.* Ultrastructural contributions to molecular pharmacology, *A. Jones and E. Mills.* Isolation and characterization of pharmacological receptors, *D. Shirachi, S. Chan, and A. Trevor.* Significance of the study of homologous series to the understanding of the relation between structure and activity, *M. Chenoweth, L. McCarty, and W. Piper.* Analgesic receptors, *A. Casy.* The d-tubocurarine receptor, *D. Taylor.* Conformational changes induced in proteins by drug molecules, *A. Levitzki.* Molecular pharmacology of acetylcholinesterase, *R. Kitz.* Enzyme kinetic approach to the analysis of structure function relationship in the evaluation of drug action, *J. Ayling.*

Part II: Intermolecual forces and the pharmacology of simple molecules, *K. Miller and E. Smith.* Function of the myoglobin molecule as influenced by anesthetic molecules, *W. Settle.* Role of the noble gas series in molecular pharmacology, *M. Powell.* Role of steroid hormones in the induction of specific proteins, *D. Martin, Jr., G. Rousseau, and J. Baxter.* Use of genetic variants in molecular pharmacology, *S. Hegeman.* NMR studies of the binding of drugs to macromolecules and cellular structures, *J. Fischer.* Molecular pharmacology, *D. Holmes.* Optical activity and protein-drug interactions, *C. Ottaway and D. Wetlaufer.* The use of molecular orbital theory in pharmacological studies, *W. Neely.* Calculation of the interaction of small molecules with large molecules, *E. Chan.* Relative usefulness of lipids and proteins as models for studies in molecular pharmacology, *D. Sears.*

FRIED *Methods of Neurochemistry*

a series edited by RAINER FRIED, *School of Medicine, Creighton University, Omaha, Nebraska*

Vol. 1   392 pages, illustrated. 1971
Vol. 2   312 pages, illustrated. 1972
Vol. 3   304 pages, illustrated. 1972
Vol. 4   352 pages, illustrated. 1973
Vol. 5   296 pages, illustrated. 1973

A thorough, well–documented, open–ended series of methods prepared by experts in their respective fields covering all important areas of the neurosciences, with emphasis placed on the fields of biochemistry and pharmacology. Serves as a reliable manual for pertinent laboratory procedures. Of special value to the bench chemist, the general neurochemist, all interested laboratory workers, and the medical scientist who wishes to initiate research in neurochemistry.

CONTENTS:

Volume 1: Purification and properties of isolated myelin, *L. Mokrasch.* Purification and assay of phospholipids, *G. Ansell and S. Spanner.* Methods of determination of catecholamines and their metabolites, *D. Sharman.* Microiontophoresis, *K. Krnjević.* Tabulation of compounds of importance to neurochemistry, *R. Fried and W. Menzies.* Biochemical screening and diagnostic procedures in mental retardation, *D. O'Brien.*

Volume 2: The subcellular fractionation of brain tissue with special reference to the preparation of synaptosomes and their component organelles, *V. P. Whittaker and L. A. Barker.* The effect of short-term "learning" experiences on the incorporation of radioactive precursors into RNA and polysomes of brain, *J. E. Wilson and E. Glassman.* Biochemical methods used to study single neurons of APLYSIA CALIFORNICA (sea hare), *R. P. Peterson.* Fluorometric analysis of 5-hydroxytryptamine and related compounds, *R. P. Maikel.* Tryptophan-5-hydroxylase, *E. M. Gál and S. A. Millard.* Methods for the measurement of cyclic AMP in brain, *H. Shimizu and J. W. Daly.* Assay of pyridoxal and its derivatives, *Y. H. Loo and W. M. Cort.* Current approaches in the study of receptors in the CNS, *S. G. A. Alivisatos and P. K. Seth.*

Volume 3: Assay of γ-aminobutyric acid and enzymes involved in its metabolism, *C. F. Baxter.* Isolation, purification, and assay of fatty acids and steroids from the nervous system, *Y. Kishimoto and M. Hoshi.* Cerebral cortex slices and synaptosomes. *In vitro* approaches to brain metabolism, *H. F. Bradford.* The isolation and assay of the nerve growth factor proteins, *S. Varon, J. Nomura, J. R. Pérez-Polo, and E. Shooter.* Investigation of behavior induction by injection of mammalian brain extract, *A. M. Golub.*

Volume 4: Culture of nerve tissue, *J. Schneider.* Measurement of choline (CH), acetylcholine (ACH), and their metabolism by combined enzymatic and radiometric techniques, *J. Saelens, J. Simke, M. Allen, and C. Conroy.* The giant axon of the squid: A useful preparation for neurochemical and pharmacological studies, *P. Rosenberg.* Behavior techniques for experiments in behavior, learning, and memory, *R. Simon and P. Freedman.* Neurophysiological methods and models for neurochemists, *L. Massopust, Jr., L. Wolin, S. Kadoya, R. White, and N. Taslitz.*

Volume 5: Prenatal diagnosis of genetic disorders leading to mental retardation, *S. Melancon and H. Nadler.* Behavioral methods in pharmacology, *L. Seiden.* Microtubules and microtubular protein, *F. Samson, Jr., D. Redburn, and R. Himes.* Ion transport in the synaptosome and the Na+–K+–ATPase, *A. Abdel-Latif.* Critical evaluation and implications of denervation and re–innervation studies of cross–striated muscle, *E. Gutmann.*

GOODFRIEND, SEHON, and ORANGE *Control Mechanisms in Allergy*

(Immunology Series, Volume 1)

edited by L. GOODFRIEND, *Royal Victoria Hospital, Montreal, Quebec,* A. H. SEHON, *Department of Immunology, University of Manitoba, Winnipeg,* and R. P. ORANGE, *Hospital for Sick Children at the University of Toronto, Ontario*

in preparation. 1973

Covers the essential areas of allergy: cellular synthesis of antibodies; the genetics of formation in animals and humans; the relationship of paramatisia, and the interaction of these antibodies at the target cells. Also includes the chemistry and pharmacology of these target cells. Directed to immunologists and clinicians, allergists, pharmacologists, geneticists, and pediatricians.

HIMWICH *Biochemistry of the Developing Brain*

edited by WILLIAMINA A. HIMWICH, *Galesburg State Research Hospital, Illinois*

Part I   320 pages, illustrated. 1973
Part II 368 pages, illustrated. 1973

Covers biochemistry of the developing brain, including the most current ideas of leaders in the field and the progress that has been made in the last twenty years. Directed to graduate students and research workers in neurology, neurological chemistry and biology, pharmacology, biochemistry, physiology, and related areas.

CONTENTS:

Part I: History of the biochemistry of the developing brain, *H. Himwich.* Amino acids and proteins, *J. Davis and W. Himwich.* Development and regulation of the γ-aminobutyric acid system during ontogeny, *B. Haber and K. Kuriyama.* Myelination and amino acid imbalance in the developing brain, *H. Agrawal and A. Davison.* Development of amino acid transport systems in incubated brain tissue, *G. Levi.* Metabolic compartmentation in developing brain, *S. Berl.* Effects of hormones on the maturation of the brain, *R. Balazs and D. Richter.*

Part II: Some hormonal effects on brain composition, *E. Howard.* Carbohydrate metabolism in the developing brain, *J. O'Neill.* Ontogenetic aspects of cerebral metabolic relationships *in vitro*, *V. Sutherland, D. Shirachi, and A. Trevor.* Enzymes in the developing brain, *J. Van Den Berg.* Cellular growth during normal and abnormal development, *M. Winick.* Changes in the biochemical properties of the brain under the prolonged influence of hypoxia and hypokinesia, *Z. Barbashova and L. Simanovskii.*

HIRTZ *Analytical Metabolic Chemistry of Drugs*

(Medicinal Research Series, Volume 4)

by JEAN L. HIRTZ, *Ciba-Geigy Biopharmaceutical Research Center, Rueil-Malmaison, France*

translation edited by EDWARD R. GARRETT

416 pages, illustrated. 1971

Surveys the relevant literature and outlines the methods and techniques that have been used to identify and determine the products of drug metabolism. Stresses analytical procedures which isolate, characterize, and quantitatively determine minute amounts of metabolites and unchanged drugs in complex biological fluids. An invaluable reference for all those engaged in the study of drug metabolism—biochemists, toxicologists, pharmacologists, biopharmaceutical scientists, and hospital pharmacists. "Altogether a most useful book, truly what T. H. Huxley called one of the counters of science."—*The Analyst.*

CONTENTS: Phenolic acids and derivatives • Amines • Aminophenols, catecholamines • Phenothiazines • Dibenzazepines and benzodiazepines • Carbamates • Anilides • Barbiturates • Derivatives of urea, guanidine, and others • Sulfonamides • Imides • Hydrazides and hydrazines • Heterocycles with one nitrogen • Heterocycles with two nitrogens • Heterocycles with more than two nitrogens • Heterocycles not containing nitrogen • Alkaloids • Antibiotics • Glycosides • Miscellaneous drugs.

JACOB, ROSENBAUM, and WOOD *Dimethyl Sulfoxide*

In 2 Volumes

edited by STANLEY JACOB, *University of Oregon Medical School, Portland,* EDWARD ROSENBAUM, and DON C. WOOD, *Providence Hospital, Portland, Oregon*

Vol. 1 *Basic Concepts of DMSO*
496 pages, illustrated. 1970

Vol. 2 in preparation. 1974

Introduces the scientific community to some of the properties and potential uses of DMSO, and demonstrates the wide range of scientific and professional persons working with this compound. Provides the most complete bibliography of DMSO ever published. Directed to physicians, pharmacologists, veterinarians, dentists, bacteriologists, biochemists, and organic chemists.

CONTENTS: The chemistry of DMSO, *H. Szmant.* The pharmacology of DMSO, *S. Jacob.* The toxicology of DMSO in animals, *M. Mason.* Fate and metabolism of DMSO, *D. Wood.* Radioprotective and cryoprotective properties of DMSO, *M. Ashwood-Smith.* The use of DMSO in enzyme catalyzed reactions, *D. Rammler.* The role of DMSO in microbiology and serology, *G. Pottz, J. Rampey, and F. Benjamin.* DMSO in experimental immunology, *H. Raettig.* The interaction of DMSO and alcohol, *H. Mallach.* DMSO and plants, *R. Weintraub.* DMSO in veterinary medicine, *L. Koger.* Some effects of DMSO on connective tissue, *G. Gries.* DMSO in dermatology, *T. Cortese, Jr.* DMSO Bibliography.

NOTARI *Biopharmaceutics and Pharmacokinetics: An Introduction*

by ROBERT E. NOTARI, *The Ohio State University, Columbus*

336 pages, illustrated. 1971

Presents an introduction to the principles and applications of biopharmaceutics and pharmacokinetics as employed in research and clinical practice. Includes examples and practice problems for each topic and may readily be used as a textbook for students in pharmacy, pharmacology, medicinal chemistry, medicine, dentistry, and drug-oriented teaching programs. Also of value to research scientists and biochemical practitioners.

CONTENTS: Bioavailability • Rate processes in biological systems S_s • Principles of pharmacokinetics • Biopharmaceutics: Clinical applications of pharmacokinetic parameters • Additional practice problems for comprehensive review.

PÉREZ-TAMAYO and ROJKIND
Molecular Pathology of Connective Tissues

(The Biochemistry of Disease Series, Volume 3)

edited by RUY PÉREZ-TAMAYO, *Department of Cell Biology, National University of Mexico, Mexico City,* and MARCOS ROJKIND, *Department of Cell Biology, National Polytechnic Institute, Mexico City, Mexico*

408 pages, illustrated. 1973

Reviews several aspects of current basic research in connective tissues and systematically relates them to disease. Emphasizes the known or suspected relations of recently acquired information to specific pathological processes. Of fundamental concern to biochemists, pathologists, geneticists, physicians, medical students, and all others involved in medical research.

CONTENTS: Molecular structures of the fibrous components of the connective tissue, *M. Rojkind.* Biosynthesis of collogen and elastin, *M. Rojkind and M. Zeichner.* Lathyrism, *C. Levene.* Collagen degradation and resorption physiology and pathology, *R. Pérez–Tamayo.* Biosynthesis of mucopolysaccharides and protein–polysaccharides, *J. Silbert.* Catabolism of mucopolysaccharides and protein polysaccharides, *J. Silbert.*

RUBIN *Search for New Drugs*

(Medicinal Research Series, Volume 6)

edited by ALAN A. RUBIN, *Endo Laboratories, Inc., Garden City, New York*

464 pages, illustrated. 1972

Provides information on the current status of various classes of drugs and analyzes the developmental capabilities of selected areas of drug research. Devoted to delineating areas of methodological and conceptual importance, thus acting as an important reference tool for the investigator. A treatise of value to chemists and biologists engaged in research on new and more effective therapeutic agents.

CONTENTS: Drugs for chronic inflammatory disease, *H. Paulus and M. Whitehouse.* Gastrointestinal disorders—peptic ulcer disease, *J. Thompson.* Psychoactive drug evaluation: Philosophy and methods, *S. Irwin.* Selective beta receptor drug interactions, *B. Levy.* Drugs affecting atherosclerosis, *D. Kritchevsky.* The interferon system, *R. Adamson, H. Levy, and S. Baron.* Antithrombotic and thrombolytic drugs, *K. Moser.* Pharmacolongevity: Control of aging by drugs, *F. LaBella.* Drugs affecting facilitation of learning and memory, *W. Essman.*

OTHER BOOKS OF INTEREST

BASERGA *The Cell Cycle and Cancer*

(The Biochemistry of Disease Series, Volume 1)

edited by RENATO BASERGA, *Temple University, Philadelphia, Pennsylvania*

496 pages, illustrated. 1971

BEKEY and SCHWARTZ
Hospital Information Systems

(Biomedical Engineering Series, Volume 1)

edited by GEORGE A. BEKEY, *University of Southern California, Los Angeles,* and MORTON D. SCHWARTZ, *California State College, Long Beach*

416 pages, illustrated. 1972